Interior textiles

The Textile Institute and Woodhead Publishing

The Textile Institute is a unique organisation in textiles, clothing and footwear. Incorporated in England by a Royal Charter granted in 1925, the Institute has individual and corporate members in over 90 countries. The aim of the Institute is to facilitate learning, recognise achievement, reward excellence and disseminate information within the global textiles, clothing and footwear industries.

Historically, The Textile Institute has published books of interest to its members and the textile industry. To maintain this policy, the Institute has entered into partnership with Woodhead Publishing Limited to ensure that Institute members and the textile industry continue to have access to high calibre titles on textile science and technology.

Most Woodhead titles on textiles are now published in collaboration with The Textile Institute. Through this arrangement, the Institute provides an Editorial Board which advises Woodhead on appropriate titles for future publication and suggests possible editors and authors for these books. Each book published under this arrangement carries the Institute's logo.

Woodhead books published in collaboration with The Textile Institute are offered to Textile Institute members at a substantial discount. These books, together with those published by The Textile Institute that are still in print, are offered on the Woodhead website at: www.woodheadpublishing.com. Textile Institute books still in print are also available directly from the Institute's website at: www.textileinstitutebooks.com.

A list of Woodhead books on textile science and technology, most of which have been published in collaboration with The Textile Institute, can be found towards the end of the contents pages.

Woodhead Publishing in Textiles: Number 92

Interior textiles

Design and developments

Edited by
T. Rowe

CRC Press
Boca Raton Boston New York Washington, DC

WOODHEAD PUBLISHING LIMITED
Oxford Cambridge New Delhi

Published by Woodhead Publishing Limited in association with The Textile Institute
Woodhead Publishing Limited, Abington Hall, Granta Park, Great Abington
Cambridge CB21 6AH, UK
www.woodheadpublishing.com

Woodhead Publishing India Private Limited, G-2, Vardaan House, 7/28 Ansari
Road, Daryaganj, New Delhi – 110002, India
www.woodheadpublishingindia.com

Published in North America by CRC Press LLC, 6000 Broken Sound Parkway,
NW, Suite 300, Boca Raton, FL 33487, USA

First published 2009, Woodhead Publishing Limited and CRC Press LLC

British Library Cataloguing in Publication Data
A catalogue record for this book is available from the British Library.

Library of Congress Cataloging in Publication Data
A catalog record for this book is available from the Library of Congress.

Woodhead Publishing ISBN 978-1-84569-351-0 (book)
Woodhead Publishing ISBN 978-1-84569-687-0 (e-book)
CRC Press ISBN 978-1-4200-9482-4
CRC Press order number WP9482

The publishers' policy is to use permanent paper from mills that operate a sustainable forestry policy, and which has been manufactured from pulp which is processed using acid-free and elemental chlorine-free practices. Furthermore, the publishers ensure that the text paper and cover board used have met acceptable environmental accreditation standards.

Typeset by SNP Best-set Typesetter Ltd., Hong Kong
Printed by TJ International Limited, Padstow, Cornwall, UK

Contents

Contributor contact details

(* = main contact)

Editor

Mr Trevor Rowe
The University of Bolton
Deane Road
Bolton
BL3 5AB
UK

Email: tr2@bolton.ac.uk

Chapter 1

Professor Dr R. Kozlowski*, A. Kicińska-Jakubowska, and M. Muzyczek
Institute of Natural Fibers and Medicinal Plants
71 B Wojska Polskiego Street
60-630 Poznan
Poland

E-mail: iza.maciejowska@inf.poznan.pl; ryszard.kozlowski@esscorena.net

Chapter 2

Professor Dr R. Milašius* and Dr V. Jonaitienė
Department of Textile Technology
Kaunas University of Technology
Studentu St. 56
LT-51424
Kaunas
Lithuania

E-mail: rimvydas.milasius@ktu.lt

Chapter 3

Professor S. J. Kadolph
Department of Apparel, Educational Studies, and Hospitality Management
31 MacKay Hall
Iowa State University
Ames
IA 50011-1121
USA

E-mail: skadolph@iastate.edu

Chapter 4

Trina Das
Sr. Designer
Dicitex Decor Pvt. Ltd
1081 Solitaire Corporate Park, Building No. 10
167 Guru Hargovindji Marg
Andheri East
Mumbai 400093
India

E-mail: dastrina@gmail.com

Chapter 5

D. Whitefoot
The Carpet Foundation
MCF Complex
60 New Road
Kidderminster
DY10 1AQ
UK

E-mail: d.whitefoot1@btinternet.com

Chapter 6

Professor Dr E. Strazdiene
Department of Clothing and Polymer Products Technology
Kaunas University of Technology
Studentu St. 56
LT-51424
Kaunas
Lithuania

E-mail: eugenija.strazdiene@ktu.lt

Chapter 7

Lisa M. Tucker, PhD
AIA, ASID, IIDA, LEED AP
Assistant Professor
School of Architecture + Design
Virginia Tech
201 Cowgill Hall (0205)
Blacksburg
VA 24061
USA

E-mail: ltucker@vt.edu

Chapter 8

Dr Linda Nussbaumer, PhD, CID, ASID, IDEC
Professor and Assistant Department Head
Design, Merchandising, & Consumer Sciences
South Dakota State University
Box 2275A; SNF 216
Brookings, SD 57007-0496
USA

E-mail: linda.nussbaumer@sdstate.edu

Chapter 9

Stefan Posner
Swerea IVF
Postbox 104
SE-431 22 Mölndal
Sweden

Email: stefan.posner@swerea.se

Chapter 10

Dr Charles Martin Fleischmann, PhD, PE
Associate Professor
Civil and Natural Resources Engineering
University of Canterbury
67 Creyke Road
Christchurch
New Zealand

E-mail: charles.fleischmann@canterbury.ac.nz

Chapter 11

Professor Dr Alexander Büsgen
Niederrhein University of Applied Sciences
Department of Textile and Clothing
Webschulstr. 31 41065
Mönchengladbach
Germany

Email: Alexander.Buesgen@hs-niederrhein.de

Woodhead Publishing in Textiles

1. **Watson's textile design and colour Seventh edition**
   Edited by Z. Grosicki
2. **Watson's advanced textile design**
   Edited by Z. Grosicki
3. **Weaving Second edition**
   P. R. Lord and M. H. Mohamed
4. **Handbook of textile fibres Vol 1: Natural fibres**
   J. Gordon Cook
5. **Handbook of textile fibres Vol 2: Man-made fibres**
   J. Gordon Cook
6. **Recycling textile and plastic waste**
   Edited by A. R. Horrocks
7. **New fibers Second edition**
   T. Hongu and G. O. Phillips
8. **Atlas of fibre fracture and damage to textiles Second edition**
   J. W. S. Hearle, B. Lomas and W. D. Cooke
9. **Ecotextile '98**
   Edited by A. R. Horrocks
10. **Physical testing of textiles**
    B. P. Saville
11. **Geometric symmetry in patterns and tilings**
    C. E. Horne
12. **Handbook of technical textiles**
    Edited by A. R. Horrocks and S. C. Anand
13. **Textiles in automotive engineering**
    W. Fung and J. M. Hardcastle
14. **Handbook of textile design**
    J. Wilson
15. **High-performance fibres**
    Edited by J. W. S. Hearle
16. **Knitting technology Third edition**
    D. J. Spencer
17. **Medical textiles**
    Edited by S. C. Anand

18 **Regenerated cellulose fibres**
Edited by C. Woodings

19 **Silk, mohair, cashmere and other luxury fibres**
Edited by R. R. Franck

20 **Smart fibres, fabrics and clothing**
Edited by X. M. Tao

21 **Yarn texturing technology**
J. W. S. Hearle, L. Hollick and D. K. Wilson

22 **Encyclopedia of textile finishing**
H-K. Rouette

23 **Coated and laminated textiles**
W. Fung

24 **Fancy yarns**
R. H. Gong and R. M. Wright

25 **Wool: Science and technology**
Edited by W. S. Simpson and G. Crawshaw

26 **Dictionary of textile finishing**
H-K. Rouette

27 **Environmental impact of textiles**
K. Slater

28 **Handbook of yarn production**
P. R. Lord

29 **Textile processing with enzymes**
Edited by A. Cavaco-Paulo and G. Gübitz

30 **The China and Hong Kong denim industry**
Y. Li, L. Yao and K. W. Yeung

31 **The World Trade Organization and international denim trading**
Y. Li, Y. Shen, L. Yao and E. Newton

32 **Chemical finishing of textiles**
W. D. Schindler and P. J. Hauser

33 **Clothing appearance and fit**
J. Fan, W. Yu and L. Hunter

34 **Handbook of fibre rope technology**
H. A. McKenna, J. W. S. Hearle and N. O'Hear

35 **Structure and mechanics of woven fabrics**
J. Hu

36 **Synthetic fibres: nylon, polyester, acrylic, polyolefin**
Edited by J. E. McIntyre

37 **Woollen and worsted woven fabric design**
E. G. Gilligan

38 **Analytical electrochemistry in textiles**
P. Westbroek, G. Priniotakis and P. Kiekens

39 **Bast and other plant fibres**
R. R. Franck

40 **Chemical testing of textiles**
Edited by Q. Fan
41 **Design and manufacture of textile composites**
Edited by A. C. Long
42 **Effect of mechanical and physical properties on fabric hand**
Edited by Hassan M. Behery
43 **New millennium fibers**
T. Hongu, M. Takigami and G. O. Phillips
44 **Textiles for protection**
Edited by R. A. Scott
45 **Textiles in sport**
Edited by R. Shishoo
46 **Wearable electronics and photonics**
Edited by X. M. Tao
47 **Biodegradable and sustainable fibres**
Edited by R. S. Blackburn
48 **Medical textiles and biomaterials for healthcare**
Edited by S. C. Anand, M. Miraftab, S. Rajendran and J. F. Kennedy
49 **Total colour management in textiles**
Edited by J. Xin
50 **Recycling in textiles**
Edited by Y. Wang
51 **Clothing biosensory engineering**
Y. Li and A. S. W. Wong
52 **Biomechanical engineering of textiles and clothing**
Edited by Y. Li and D. X-Q. Dai
53 **Digital printing of textiles**
Edited by H. Ujiie
54 **Intelligent textiles and clothing**
Edited by H. Mattila
55 **Innovation and technology of women's intimate apparel**
W. Yu, J. Fan, S. C. Harlock and S. P. Ng
56 **Thermal and moisture transport in fibrous materials**
Edited by N. Pan and P. Gibson
57 **Geosynthetics in civil engineering**
Edited by R. W. Sarsby
58 **Handbook of nonwovens**
Edited by S. Russell
59 **Cotton: Science and technology**
Edited by S. Gordon and Y-L. Hsieh
60 **Ecotextiles**
Edited by M. Miraftab and A. Horrocks
61 **Composite forming technologies**
Edited by A. C. Long

83 **Smart clothes and wearable technology**
Edited by J. McCann and D. Bryson

84 **Identification of textile fibres**
Edited by M. Houck

85 **Advanced textiles for wound care**
Edited by S. Rajendran

86 **Fatigue failure of textile fibres**
Edited by M. Miraftab

87 **Advances in carpet technology**
Edited by K. Goswami

88 **Handbook of textile fibre structure Volumes 1 and 2**
Edited by S.J. Eichhorn, J.W.S. Hearle, M. Jaffe and T. Kikutani

89 **Advances in knitting technology**
Edited by T. Dias

90 **Smart textile coatings and laminates**
Edited by W.C. Smith

91 **Handbook of tensile properties of textile and technical fibres**
Edited by A.R. Bunsell

92 **Interior textiles: Design and developments**
Edited by T. Rowe

93 **Textiles for cold weather apparel**
Edited by J.T. Williams

94 **Modelling and predicting textile behaviour**
Edited by X. Chen

95 **Textiles for construction**
Edited by G. Pohl

96 **Engineering apparel fabrics and garments**
J. Fan and L. Hunter

97 **Surface modification of textiles**
Edited by Q. Wei

Part I

Fundamental principles of interior textiles

1

Natural fibres for interior textiles

R. KOZLOWSKI, A. KICIŃSKA-JAKUBOWSKA, M. MUZYCZEK, Institute of Natural Fibres, Poland

Abstract: Natural fibres used in interior finishing are of great importance as their properties have positive influence on users. The chapter provides characteristics of natural fibres and the main areas of their application in interiors. The authors emphasize their potential barrier functions and effect on human well-being. The chapter also discusses advantages and disadvantages of using these fibres in interiors. It also lists key organizations and sources of information on natural fibres.

Key words: natural fibres classification, interior furnishing, bedding and mattresses, upholstery composites.

1.1 Introduction: Basic principles of fibres, yarns and fabrics

Natural fibres have played a part in human existence for thousands of years. They are highly valuable textile raw materials characterized by specific features that are user-friendly to both people and the natural environment. Nowadays, due to their specific properties, they find applications in almost all branches of the economy and serve as excellent raw materials for the manufacturing of so-called 'green products'. Most natural fibrous materials are produced as sustainable and renewable resources and are generally obtained by photosynthesis. The complete cycle of their production and recycling is demonstrated in Fig. 1.1. Currently natural fibres are produced nearly all over the world and constitute a fairly significant part of international trade. The area of their applications is still growing and widening.

Figure 1.2 shows that natural fibres constitute about half of all global fibre production. Production of some natural fibres is presented in Fig. 1.3 and worldwide fibre consumption in Fig. 1.4. The world fibre market has enjoyed growth in almost every sector. While the usage of cotton, wool and silk in 2005/6 increased by 4.7% to 27.9 million tonnes, that of man-made fibres rose by 3.4% to 40.8 million tonnes. (Another sector, that of kapok, ramie, flax, hemp, jute, sisal and coir, is estimated to have stagnated at 5.8 million tonnes.)

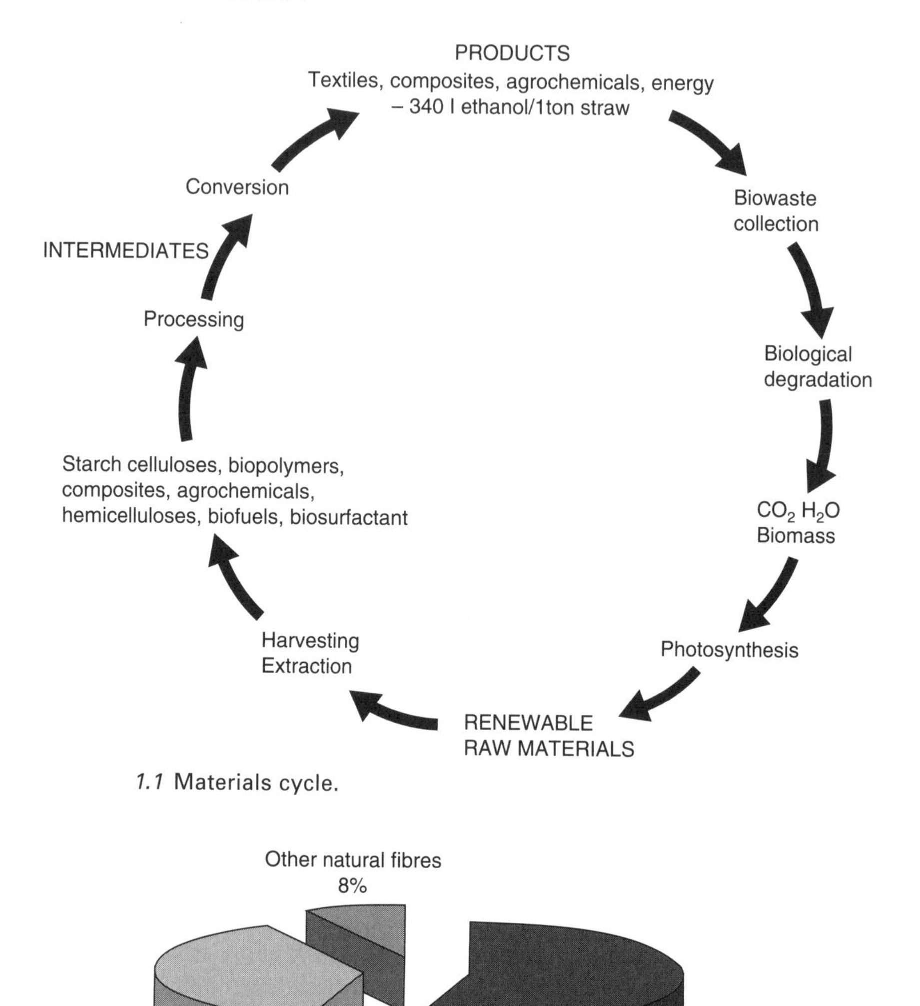

1.1 Materials cycle.

1.2 Production scale of natural fibres compared with other fibres. Source: Sauer Report 'The Fiber Year 2006/2007'.

Figure 1.5 shows global fibre production. Demand for cotton, wool, silk and man-made fibre has increased by 3.9% to 68.7 million tonnes, slightly above the average annual long-term growth rate of 3.3% (1980–2006).

Figure 1.5 also shows the long-term inter-fibre competition. Since the beginning of the 1990s, man-made fibres have been the most important

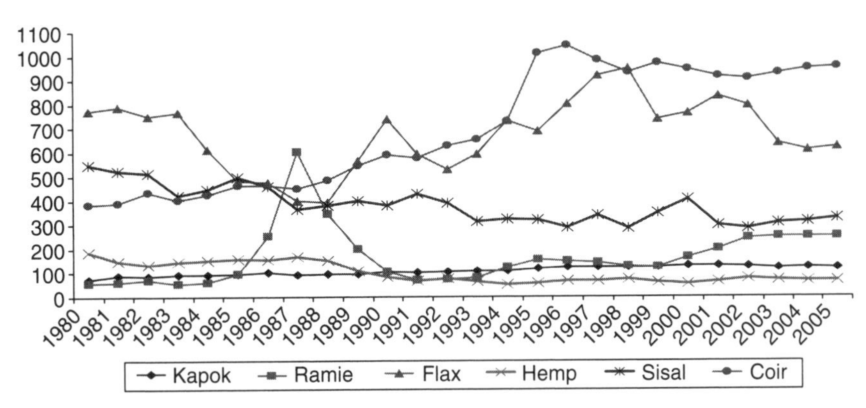

1.3 Worldwide production of some natural fibres (1980–2005), kapok, ramie, flax, hemp, sisal, coir.
Source: Sauer Report 'The Fiber Year 2005/2006'.

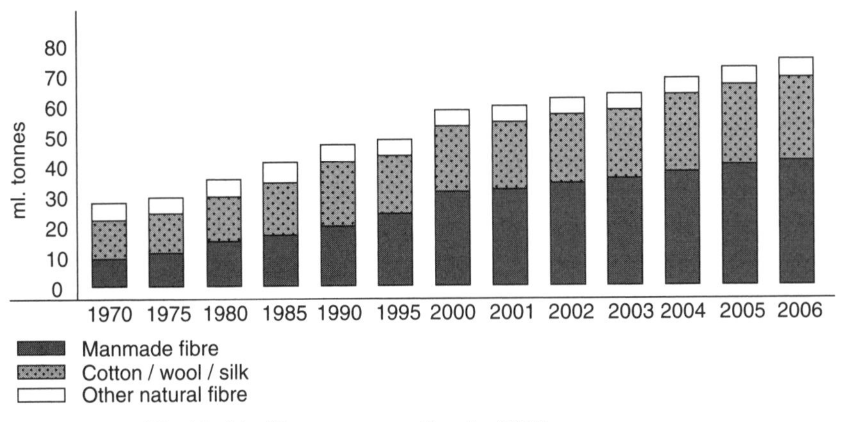

1.4 Worldwide fibre consumption in 2006.
Source: Sauer Report 'The Fiber Year 2006/2007'.

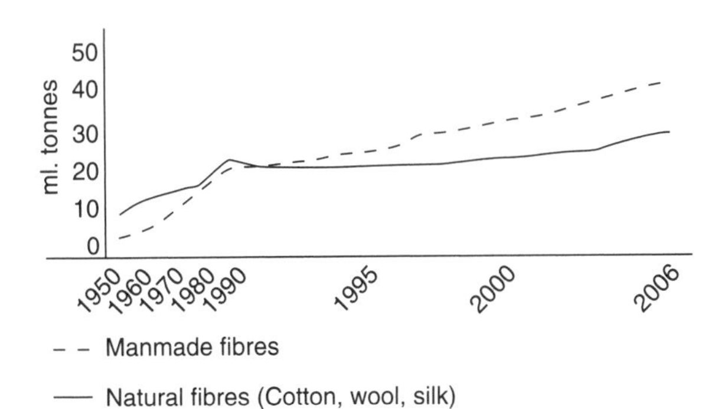

1.5 Global fibre production.
Source: Sauer Report 'The Fiber Year 2006/2007'.

fibre type in terms of volume. The average annual growth rate since 1980 for man-made fibres has been 4.1%, while for natural fibres it has been 2.4%.

Products made of natural fibres guarantee optimal comfort of use and exert a positive influence on human physiology. By direct contact with skin and the air which is then inhaled inside buildings, they minimize detrimental effects on the people living there. The nerve endings in the body collect stimuli from the skin and breathing system and finally send information to the brain. This is main reason that the interior environment influences human feelings of either comfort or discomfort. The latest research shows that natural fibres, if they are in close contact with the human body, have a positive effect on reducing oxidative stress, muscle tension and sleeping disorders [19,20].

1.2 Main types and characteristics of natural fibres as raw materials for the production of natural textiles

Natural fibres can be classified into those of animal, vegetable and mineral origin. Figure 1.6 illustrates this classification [7,23]. Tables 1.1 and 1.2 show the specific characteristics of natural fibres.

The main component of plant fibres is cellulose $(C_6 H_{10} O_5)_n$. The cellulose that is most free from encrusting substances is found in cotton and ramie fibres (83 to 99%). The higher the cellulose content and the higher its purity, the better and more valuable the fibre is. Apart from cellulose, plant fibres contain other substances that accumulate in middle plates and fibre walls, such as lignins, pectins and hemicellulose. The composition of several plant fibres is shown in Table 1.3.

1.2.1 Plant fibres and textiles

*Abaca (*Musa Textilis Nee*)*

Abaca fibre is obtained from an evergreen, perennial tropical plant called fibrous banana. The fibre is often known as *Manila*. The plant is commonly cultivated in the Philippines, and also on Java, Sumatra, and Borneo and in the countries of Central and South America. In order to obtain the fibre, the plant trunks are cut as low as possible. Fibre extraction should take place within 48 hours of cutting the plant, otherwise the fibre strength decreases and its colour changes. The fibre is used for plaiting, thick fabrics, fishing nets, sails, ship ropes, paper, tea and coffee bags, disposable fabrics and boards used in construction. The diameter of elementary fibres is about 10–30 μm.

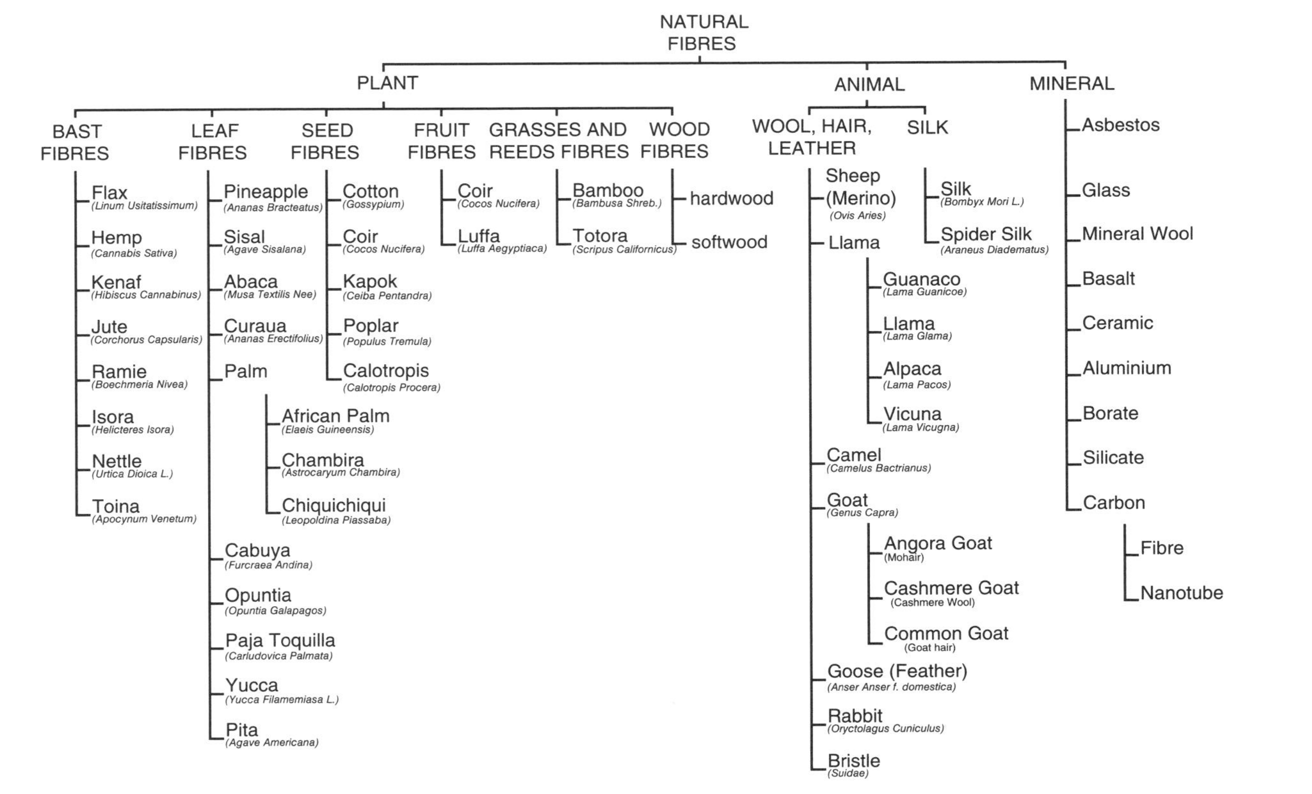

1.6 Classfication of natural fibres.
Source: Institute of Natural Fibres and Textile Terms and Definitions, The Textile Institute.

Table 1.1 Diameter of elementary fibre [μm]

Fibre	Diameter of elementary fibre (μm)
Abaca (*Musa Textilis Nee*)	10–30
African Palm (*Elaeis Guineensis*)	14–28
Alpaca (*Lama Pacos*)	15–45
Angora Goat (*Capra Hircus Aegagrus*)	24–40
Cabuya (*Furcraea Andina*)	23–35
Calotropis (*Calotropis Procera*)	20–40
Chambira (*Astrocaryum Chambira*)	8–16
Chiquichiqui (*Leopoldinia Piassaba*)	10–20
Camel (*Camelus Bactrianus*)	18–40 and 50–100
Coir (*Cocos Nucifera*)	10–16
Cotton (*Gossypium*)	15–24
Curaua (*Ananas Erectifolius*)	7–10
Flax (*Linum Usitatissimum*)	15–22
Goose Feather (*Anser Anser f. domestica*)	4.5–15
Hemp (*Cannabis Sativa*)	17–24
Isora (*Helicteres Isora*)	10–17
Jute (*Corchorus Capsularis*)	15–35
Kenaf (*Hibiscus Cannabinus*)	13–20
Lama (*Lama Glama*)	12–50
Luffa (*Luffa Aegyptiaca*)	20–50
Nettle (*Urica Dioica L.*)	25–50
Opuntia (*Opuntia Galapagos*)	20–45
Paja Toquilla (*Carludovica Palmata*)	10–20
Palmyra (*Borassus Flabelliformis*)	12–25
Pineapple Fibre (*Ananas Bracteatus*)	25–34
Pita (*Agave Americana*)	12–18
Poplar (*Populus Tremula*)	6–22
Ramie (*Boehmeria Nivea*)	30–80
Sheep's Wool (*Ovis Aries*)	20–30
Silk (*Bombyx Mori L*)	10–25
Sisal (*Agave Sisalana*)	15–30
Spider Silk (*Araneus Diadematus*)	2–8
Totora (*Scirpus Californicus*)	6–16
Vicuna (*Vicugna vicugna*)	12–15

Source: Research work INF, Poznań.

*African Palm (*Elaeis Guineensis*)*

The African Palm (*Elaeis Guineensis*) is native to West Africa, occurring between Angola and Gambia. The world's largest producer and exporter of palm oil today is Malaysia, producing about 47% of the world's supply. Indonesia is the second largest, producing approximately 36% of the world's palm oil volume. Mature trees are single-stemmed, and grow to 20 m tall. The leaves reach between 3 and 5 m in length. This kind of palm

Table 1.2 Tensile strength of bundle fibres [cN/tex]

Fibre	Tensile strength of bundle fibres (cN/tex)
Abaca (*Musa Textilis Nee*)	67
African Palm (*Elaeis Guineensis*)	23
Cabuya (*Furcraea Andina*)	40
Curaua (*Ananas Erectifolius*)	23
Flax (*Linum Usitatissimum*)	40
Hemp (*Cannabis Sativa*)	30
Jute (*Corchorus Capsularis*)	36
Kenaf (*Hibiscus Cannabinus*)	41
Sisal (*Agave Sisalana*)	62
Tussah silk	35
Wool	10

Source: Research work INF, Poznań.

Table 1.3 Chemical composition (%) of plant fibres

Fibre	Cellulose (%)	Lignin (%)	Pectin and hemicelluloses (%)
Abaca	60–80	6–14	13
Bamboo	26–43	21–31	–
Cabyua	80	17	–
Coir	36–43	41–45	3–4
Cotton	83–99	6	5
Curaua	70–80	13	–
Flax	64–84	0.6–5	19
Hemp	67–78	3.5–5.5	17
Henequen	60–78	8–13	4–28
Isora	75	23	–
Jute	51–78	10–15	37
Kenaf	44–57	15–19	–
Nettle	53–82	0.5	0.9–4.8
Pineapple	80	13	–
Pita	80	17	–
Ramie	67–99	0.5–1	22
Sisal	60–80	6–14	13

Source: [4,10,12,17].

is grown for its clusters of fruit, which can weigh 40–50 kg. Each fruit contains a single seed or nut (the palm kernel) surrounded by a soft, oily pulp. The seeds are used for the production of soap and edible vegetable oil. The fibres are extracted from the seeds. The diameter of the elementary fibres is 14–28 μm. Annually, each hectare of oil palm, which is harvested all year round, produces on average 10 tonnes of fruit yielding

3000 kg of pericarp oil, and 750 kg of seed kernels yielding 250 kg of high-quality palm kernel oil as well as 500 kg of kernel meal.

Apocynum Venetum

Apocynum Venetum is a perennial plant from the *Apocynaceae* family, which grows wild in central Asia, Siberia and Eastern Europe. The family comprises about 1500 species. The length of an elementary fibre varies between 4 and 55 mm. The fibre extracted from *Apocynum venetum* is a good raw material for obtaining fibre resembling cotton. It is characterized by a high breaking tenacity and springiness and, depending on the technology used for pre-treatment, can be used for the production of fabrics of different thicknesses, fabrics for bags, and ropes and in pulp and paper production. However, its stems cause problems in retting because of the high content of resins and tannins [12].

*Bamboo (*Bambusa Shreb.*)*

Bamboo (*Bambusa Shreb.*) is a perennial plant that grows in monsoon climates, forming forests up to 40 m in height. It is used in applications such as construction, carpentry, weaving and plaiting. New developments include towels made of bamboo; since bamboo contains a hollow thread, it is particularly good at absorbing moisture. Bamboo fibres used for interior decorating provide antibiosis, bacteriostatic and deodorization functions and protection against ultraviolet radiation. The sun's ultraviolet radiation, heat and light are major contributors to the fading and deterioration of fabrics, carpets, furnishings, store-front displays and other interior items. Curtains made of bamboo fibre can absorb ultraviolet radiation in various wavelengths, thus they lessen the deterioration of fabrics and may help prevent the devastating effects of ultraviolet rays on the body, such as premature aging, wrinkles, sunspots and even skin cancer.

*Cabuya (*Furcraea Andina*)*

Cabuya belongs to leaf succulents or trees of the *Agavales* order. The plants are found in Central America, Mexico and the Caribbean. The diameter of elementary fibres is about 23–35 μm. The fibres are used for ropes, bags, nets, paper and also hats.

*Calotropis (*Calotropis Procera*)*

Calotropis (*Calotropis Procera*) is a delicate fibre resembling cotton, grown in Asia and Africa. The diameter of elementary fibres is 20–40 μm.

*Chambira (*Astrocaryum Chambira*)*

Chambira, from the *Palmaceae* family, is a highly valued raw material used for the production of ropes, lines, nets, hammocks, bags, brushes, mats and roofing. The fibre can also be used for diving equipment and necklace production. It has edible fruits, resembling coconuts, used for oil production. The plant is grown in many parts of Amazonia (Peru, Bolivia, Columbia and Venezuela). The tree reaches 30 m in height and the diameter of elementary fibres is 8–16 μm.

*Chiquichiqui (*Leopoldinia Piassaba*)*

Chiquichiqui (*Leopoldinia Piassaba*) is a 7 to 12 m high palm tree which grows in the wetlands of northern Brazil and Venezuela. For industrial purposes, the fibre is extracted mainly from older leaves. After cleaning, the fibre is 60 to 180 mm long and is chocolate or brown in colour, with lighter streaks. *Leopoldinia Piassaba* absorbs water easily and loses flexibility after wetting. The fibre is quite fragile and is used locally for the production of rope, string, brushes and baskets. The diameter of elementary fibres is 10–20 μm.

*Coir (*Cocos Nucifera*)*

Coir comes from coconut shells. It is grown mostly in western, central and southern Africa, and India, and on the Ivory Coast. The world production of coir fibre is estimated at 250000 tonnes. Coir fibre is reddish-brown, very hard and difficult to separate but also very resistant to mechanical stress and the effects of water. The diameter of elementary fibres is 10–16 μm. Coir is used for brushes, mattresses, bags, ropes, mats, matting, doormats, carpets, wall-panelling and handicrafts. Coir fibre is also used to cover materials such as irrigation pipes. In addition, combined with other materials, it is used for upholstery and in the automotive industry [14].

*Cotton (*Gossypium Hirsutum*)*

The most common seed fibre is cotton. It is a soft, staple fibre that grows around the seeds of the cotton plant (*Gossypium*), a shrub native to tropical and subtropical regions around the world, including the Americas, India, and Africa [22,24]. In 2007, the eleven largest producers of cotton in the world were China, India, the United States of America, Pakistan, Brazil, Australia, Uzbekistan, Turkey, Greece, Turkmenistan and Syria. These countries were responsible for about 85% of the world's cotton production.

The diameter of elementary fibres is about 15–25 μm and the length of cotton fibres is about 10–50 mm. The waxes and fats present on the surface of the fibre make it smooth and flexible. Cotton fibres are characterized by small elongation which ensures that clothing will not change its shape during use. Cotton is also used for interior products. The best examples are highly absorbent bath towels and robes, upholstery materials, household textiles such as dishcloths, dusters, bed linen, tablecloths, handicrafts, roller blinds, curtains and small pillows. Cotton is widely used in blends with other natural and man-made fibres.

Cotton can also be cultivated organically without using pesticides and chemical additives in fertilizers; cotton from organic plantations is used for the production of more luxurious products, from tissues to kimonos.

*Curaua (*Ananas Erectifolius*)*

Curaua belongs to the *Bromeliae* family and originates from South America. Its main advantages are that it is odourless and soft to the touch, with a weight that is lower than that of silk. The diameter of elementary fibres is 7–10 μm. The fibre yield is around 6–7% of the weight of the leaf. One hectare with twenty thousand plants will produce 2.4–3 tonnes per year. Due to its excellent shock-absorbing properties, curaua is used in the automotive industry for dashboards, in the electrical and electronic industries, and for the production of thin ropes, twines, threads, nets, hammocks, handicrafts and all kinds of hand-made products used in interiors, and also in apparel.

*Flax (*Linum Usitatissimum*)*

Bast is the most common plant group, with its main member being flax (*Linum usitatissimum*). Flax is one of the oldest textile fibres in the world and can be grown in a wide range of latitudes, from warm to cold climates. The diameter of elementary fibres is about 15–25 μm. Flax fibres are used for clothing, decorative fabrics, tablecloths, bed sheets, non-wovens and also ropes. Flax shives are also used for the production of particle boards, fibre for thermo-insulating materials, paper and composites [1,4]. Clothing made of linen (flax) has characteristics which make it very comfortable to wear, e.g. good air permeability, high hygroscopicity, coolness, softness, and low susceptibility to electrostatic charges on its surface, which all have a positive impact on human physiology. There properties are also very important when using flax fabrics, curtains, and tapestries for room interiors. Linen fabrics are valued by both designers and users for their smoothness, silky gloss and high durability. The lignin content in its chemical composition enables products made of this fibre to protect the user efficiently from UV radiation [28–30]. These advantages make flax fibre especially suitable for creating summer collections of clothes worn in warm climates. Short flax fibres are excellent raw materials for cigarette paper (with low tar emission when lit). Flax is used for household textiles such as tablecloths, serviettes, sheets, pillowcases, duvet covers, towels, glass and floor cloths, tea-towels, bed sheets, decorative fabrics, floor coverings, curtain fabrics, furnishing fabrics and industrial end-uses. Flax seeds are used in nutrient and pharmaceutical products [25–27].

*Hemp (*Cannabis Sativa*)*

Another bast plant is hemp (*Cannabis Sativa*), an annual green plant species which grows in moderate climatic conditions in Europe, North and South America, and Asia. Hemp fibres are characterized by high tenacity, high hygroscopicity and quick moisture transport, high strength and low homogeneity. The average length of hemp fibre is about 17–24 mm and the elementary fibre diameter is 10–17 μm. It is increasingly found in clothing, in technical products such as composites and in upholstery material, carpet underlay, non-wovens, and household textiles such as towels, bed linen and tablecloths. Due to their great ability to protect against UV radiation (lignin content), hemp fabrics should be used for the production of curtains, blinds and wall coverings. Because the yarns in hemp fabrics swell and enlarge their diameter, they are used for making roof coverings. Hemp seeds also find applications in the food (hemp-seed oil) and cosmetic industries.

*Isora (*Helicteres Isora*)*

Isora is extracted from the bark of *Helicteres Isora* during the retting process [15]. It is cultivated in many regions of Southern India. There are two notable varieties of the plant: *tomentosa* and *glabrescens*. In the former, the underside of the leaves is glabrous, and in the latter, both sides of the leaves are glabrous. Fibre present in the inner part of the bark is peeled off, washed and dried. The fibre has a pale-yellow to light-brown colour. The best type of fibre is obtained when the plants are 1–1.5 years old; plants older than 2 years yield coarse and brittle fibre. The fibre is used for the production of ropes, and for weaving boxes and bags. The diameter of elementary fibres is about 10–17 μm.

*Jute (*Corchorus Capsularis*)*

Jute (*Corchorus Capsularis*), ramie (*Boehmeria Nivea*) and isora (*Helicteres Isora*) deserve special attention. Jute fibre is extracted from the stem of an annual plant (*Carchorus* genus) belonging to the *Tiliaceae* family, grown in Asia, Europe and South America. Its technical fibre is long and soft with a lot of lustre. The diameter of elementary fibres is about 15–35 μm. In cross-section it has a polygonal shape with a canal of a different size. It is used for packaging, conveyor belts, upholstery and decorative fabrics, rugs, wall coverings and carpets. Finished jute-blended fabrics make excellent furnishing fabrics for use in upholstery, curtains and screens. Jute/PP fabrics can be made into attractive blankets after the right finishing operations, particularly raising. Jute is an excellent material for composites such as panelling, partitions, doors, window frames and furniture.

*Kenaf (*Hibiscus Cannabinus*)*

Another important bast fibre is extracted from kenaf (*Hibiscus Cannabinus*). The plant is cultivated in Asia, Africa, America and also Europe. Its stem thickness can reach 8–30 mm, the elementary fibre diameter is 13–20 μm, and the length of its technical fibre is 100–250 cm. The stems are from 1 to 5 m tall. Harvesting for fibre is done at the time of seed capsule formation. Kenaf fibre is highly resistant to water, yet it breaks easily. It is used for the production of thick wound-dressing fabrics, rope, paper and composite products.

*Luffa (*Luffa Aegyptiaca*)*

Luffa is an annual tropical creeping plant, its creepers reaching 9 metres. It is cultivated in Africa, Asia, and Australia for its cylindrical fruits (61 cm long and 7.6 cm wide). The diameter of elementary fibres is 20–50 μm. Luffa is used for the production of decorative items, boiler filters and insulating materials, for cleaning difficult surfaces and as acoustic insulation. It is used as sponges for body massaging and improving blood circulation, and for bath mats.

*Nettle (*Urica dioica L.*)*

The nettle (*Urica dioica L.*) is dioecious (separate male and female plants) and is common in moderate and cold climates. The nettle fibre is characterized by a high breaking tenacity, softness and flexibility; it is similar in appearance to ramie fibre. The length of elementary fibres varies between 4 and 620 mm. The length of the technical fibre, which is extracted either by retting the stalks or by decorticating, does not exceed 80 cm. Harvesting for fibre takes place at the stage of full flowering. Although nowadays nettle fibre is not used on an industrial scale, it was traditionally used in the production of thick fabrics, ropes and string. Currently, studies are being carried out to develop technologies that will allow for the wider use of nettle fibre in the textile industry.

*Opuntia (*Opuntia Galapagos*)*

Opuntia, from the *Cacteae* family, is also a plant which grows in North and South America, from Canada to southern Argentina, and in southern Europe. There are various species of the plant – from very small to bushes and trees. The diameter of elementary fibres is 20–45 µm. Opuntia fibre is used in construction as a solidifier and as a fibre for the reinforcing of composites.

*Paja Toquilla (*Carludovica Palmata*)*

Paja Toquilla is cultivated in South America (Ecuador). The fibre extracted from the plant is long, smooth and of high tenacity; the diameter of elementary fibres is 10–20 µm. It is mainly used for the manual plaiting of Panama hats and bags, and paper production. For weaving, hand-picked leaves are separated into smaller slivers that are soaked in water and subsequently dried in the sun. The raw material is then bleached with a sulphur compound. The main application area is in artisan production.

*Palmyra (*Borassus Flabelliformis*)*

Palmyra – a plant fibre of the *Palmaceae* family – is cultivated in India and Sri Lanka. The fibre is extracted from leaves, starting from the sixth year after planting the trees. Although the best fibre is extracted between the 8th and 10th years of cultivation, the fibre quality is determined by the region, precipitation and soil. Palmyra fibre is thick, hard and cinnamon-brown in colour. The fibre is characterized by high tenacity and stiffness; therefore it is used for the production of brushes. The diameter of elementary fibres is 12–25 µm.

*Pineapple (*Ananas Bracteatu*)*

Pineapple fibre (*Ananas Bracteatu*) comes from South America and India. Harvesting of the leaves and fruits takes place at the same time. The technical fibre is about 60–120 cm. The fibre easily absorbs water and is very resistant to abrasion. There are two methods used to extract the fibre from the leaves. The first one is extracting fibre using decorticating machines, the second one is retting leaves in water for 24–48 hours at a temperature of 15–30 °C, after which the fibres can be easily separated from the leaves. It is characterized by an intense lustre and fineness. It is used for the production of clothes, tablecloths, bags and mats. The diameter of elementary fibres is 25–34 μm.

*Pita (*Agave Americana*)*

Pita is a fibre extracted from the American agave (*Agave Americana*). The plant is grown in Mexico, Florida and the West Indies. The pita technical fibre is white or creamy with lustre, flexible and relatively light, hygroscopic and 60–200 cm long. The diameter of elementary fibres is about 12–18 μm. It is used mainly for ropes, string, paper and plaiting. Pita leaves can also be used as roofing.

*Poplar (*Populus Tremula*)*

Although the most common seed fibre is cotton, it is worth mentioning poplar (*Populus Tremula*), which grows wild in all of Europe, northern-western Africa and western Siberia. It has delicate, short fibres of diameter 6–22 μm. This use of this as a fibre has not been explored as yet. Poplar is grown for the remediation of post-mining areas and the production of wood and particle boards.

*Ramie (*Boehmeria Nivea*)*

Ramie fibre is extracted from the stem of a perennial plant belonging to the Utricaceae family, grown in Eastern Asia. The fibre length ranges from a few to more than 600 mm, and the diameter of elementary fibres is 30–80 μm. Ramie fibre is characterized by an intense lustre, good resistance to atmospheric conditions and quite high resistance to bacterial action. Ramie can be used for light clothing fabrics, technical textiles and upholstery. Dressing fabrics, ropes, fishing nets, tents, tarpaulins and fire hoses are also produced from ramie. Waste fibres are used for paper production.

*Sisal (*Agave Sizalana*)*

Sisal is one of the 300 species of the *Agave* genus, which is one of the 21 genera of the Agavacea family. Sisal and other species are cultivated in Central and South America, East Africa, Madagascar and Asia. Sisal fibre is extracted from the leaves. Extraction of the fibre should be conducted within 48 hours of cutting the leaves as with time the juice becomes glutinous, which causes difficulties in cleaning the fibre. The technical fibre in cross-section is the shape of a crescent or a horseshoe. Polygonal elementary fibres with a round canal can be seen inside. The diameter of elementary fibres is about 15–30 μm. The fibre is mostly used for ropes, string, bag fabrics, plaiting, mats, household products, carpets and other floor coverings, mattresses and composite products.

*Totora (*Scirpus Californicu*)*

Totora (*Scirpus Californicu*) is a grass fibre. It is a perennial plant that grows in the vicinity of water or marshes in South America and Easter Island. It can grow up to 4 metres in height. The diameter of elementary fibres is 6–16 μm. Totora fibre is used for rugs, wipes, seats, baskets, bags and headrests.

*Yucca (*Yucca Filamentosa L.*)*

Yucca (*Yucca Filamentosa L.*) is a perennial plant growing in subtropical and tropical climates which originated from America. Harvesting of leaves for fibre is performed once or twice a year. Both retting and chemical degumming are used for extracting the fibre. The fibre content of the leaves varies between 0.5 and 2.5 g. Yucca technical fibre resembles flax fibre and reaches 40–70 cm. It is white or yellowish, of relatively high tearing resistance, silky, ligneous and quite fragile.

1.2.2 General characteristics of various animal fibres and textiles

The most important representatives of the animal fibres are various kinds of wool, such as sheep wool (*Ovis Aries*), alpaca (*Lama Pacos*) and lama (*Lama Glama*), and silks: natural (*Bombyx Mori L*), spider (*Araneus Diadematus*) and wild (*Antheraea Pernyi*).

*Alpaca (*Lama Pacos*)*

Alpaca fibre is produced by a mammal of the *Camelidae* family [2]. The animal slightly resembles a llama, yet it is smaller and more similar to a sheep. Alpaca is kept in the Andes from southern Peru, through Chile to northern Bolivia. The species has 26 shades of colour and the alpaca hair has a silky lustre. There are two varieties of alpaca – suri and huacaya – with different types of hair. Alpaca wool is used for producing high-quality yarns and clothes, blankets, ponchos, carpets, carpeting, duvets, pillows, felt, therapeutic belts and winter hats. The diameter of elementary fibres is 15–45 μm. The goods made from alpaca fibre are light, with excellent thermal insulation.

*Angora Goat (*Capra Hircus Aegagrus*)*

Mohair, the lustrous fleece of the Angora goat (*Capra Hircus Aegagrus*), is one of the most important speciality animal fibres. It represents less than 0.02% of the total world fibre production. It is known for its unique high lustre, durability, elasticity, good insulation and extreme softness of touch. It is generally a long, straight and smooth fibre which can be dyed to deep, brilliant and fast colours [2]. According to the mean fibre fineness, mohair is generally classified into Kid mohair, Young mohair and Adult mohair. White is usually the natural colour of mohair, but brown, black or red varieties can also be found. The diameter of elementary fibres is about 24–40 μm. At present, mohair is produced in South Africa, the USA, Turkey, Argentina, Australia and New Zealand. Mohair finds applications in a wide range of apparel textiles and household textiles such as upholstery fabrics, curtains, carpets, rugs and mats, cushion covers, blankets, paint rollers, bath mats, bedcovers, bedspreads, blankets, pillowcases, duvets, mops and tapestries [2,16].

*Camel (*Camelus Bactrianus*)*

The Bactrian camel (*Camelus Bactrianus*) lives 35–40 years and within this time produces about 300–480 kg of wool. The raw material is in different forms: down of 5 cm, wool itself up to 40 cm long, and foretop with very long and thick hair – up to 60 cm. Fibre colours range from pale reddish to light brown, Chinese hair tending to be lighter in shade and finer than Mongolian hair. White fleece is the most valued but is very rare [2]. Camel wool is mostly used for blankets and knitwear, which are characterized by their softness, very good quality and excellent insulation properties.

*Goose Feather (*Anser Anser f. domestica*)*

Goose (*Anser Anser f. domestica*) and duck (*Anas domestica*) feathers are excellent materials for absorbing moisture and keeping in warmth, and they are used for filling duvets, pillows, sleeping bags and winter jackets. The best kinds of goose and duck down and feathers are characterized by very high resilience with long durability, even if used intensively. Feathers have many other applications, e.g. in composite production. Because of the hollow structure of the keratin fibres, all feathers contain a significant volume of air, which is an ideal dielectric material [63].

Leather

Leather is an important natural fibrous material which is used as much today as in ancient times. It is derived from the skin of various animals, most notably the cow. The oldest and most durable leather products came from Egypt. The wide use of leather would not be possible without knowledge of the tanning process, which allows a much better material than raw leather to be obtained. Tanning is preceded by numerous chemical and mechanical treatments. One of the most important is calcification, which enables the fibrous structure of collagen to be loosened, and the epidermis and other redundant tissues to be removed. All treatments are aimed at improving the usable features of leather and its properties [11,8].

*Llama (*Lama Glama*)*

Llama also comes from the family of *Camelidae* – it produces strong, elastic fibres with diameters of 12–50 μm. Llamas are bred in South America (most often in Peru, Bolivia and Argentina). The fibre is mainly used for the production of clothes, bed sheets, blankets, and therapeutic belts. The wool fibre covered with lanolin has special crimp and scale-like properties [2].

*Natural silk (*Bombyx Mori*)*

Natural silk (*Bombyx Mori*) is considered to be the most noble natural fibre of animal origin and is obtained from the cocoons of mulberry or oak (tasar) silkworms. Silk is the only natural fibre which exists as a continuous filament. Each *Bombyx mori* cocoon can yield up to 1600 metres of filament. The diameter of elementary fibres is about 10–25 μm. The silk fibre's triangular cross-section gives it excellent light reflection capability. China produces the most silk in the world, although silkworms are bred in Europe (Greece, Spain, Italy and Poland). Natural silk can be used for luxury textile products and carpets. Silk fabrics are characterized by their lustre, are smooth and thin, and very pleasant to touch. In the past, silk was used as the basis for paintings, especially in oriental art. It is now used for stockings, dental floss and cosmetics, and was used for parachutes (up to World War II). Tasar silk is soft and smooth, with a beautiful lustre. It is used for clothing production, fabrics such as shantung and pongee, and also for paper. Silk can absorb up to 30% of its own weight in moisture without creating a damp feeling. When moisture is absorbed, it generates 'wetting-heat', which helps to explain why silk is comfortable to wear next to the skin [2].

*Sheep wool (*Ovis Aries*)*

Sheep wool (*Ovis Aries*) – Merino is one of the best-known breeds of sheep and its wool accounts for about 30% of the total world production of wool fabric. Wool reaches a length of 40 to 125 mm, and that of high-mountain breeds can even be 150 to 400 mm, with the diameter of elementary fibres ranging between 20 and 40 μm. The largest producers of sheep wool are Australia, New Zealand, Argentina and South Africa. The fibre has many valuable properties, e.g. high hygroscopicity, absorbance of UV radiation, absorbance of sweat due to its porosity, resistance to external moisture and high abrasion resistance. Wool fibres have a natural lustre, depending on the type of wool and the climatic conditions. A characteristic feature of wool that is different from plant fibres and silk is felting, which determines the thermal properties of wool products. It is used for high-quality fabrics and knitwear for textiles, but it is mainly used for carpets, carpeting, duvets, blankets, pillows, cushion covers, pillowcases, bed covers, bedspreads, lamp covers, technical felt and sound absorbing material in high-quality loudspeakers.

*Spider Silk (*Araneus Diadematus*)*

The fibres produced from a spider's web can replace the most durable materials used presently. The method that a spider employs to produce a web is very economical. Liquid protein, which forms the web, is produced by spinning glands. It is drawn from the tiny holes at the back of the abdomen and sets in the air.

The unusual properties of spiders' webs are the result of the special structure of their internal organs. Due to a complex net of tiny glands, the spider is able to continuously control the protein composition of the thread. Moreover, by choosing the time of the day, the spider controls the temperature of the spinning process, and can also change the thickness, elasticity and structure of the yarn by releasing it more slowly or more quickly.

Spider silk is very thin; its Diameter is 2–8 μm, yet it is five times stronger than steel, and even stronger than Kevlar, but at the same time more flexible than nylon – such a combination is found nowhere else in any fibre. Spider silk is used in optical instruments. Most experiments have looked at two species, *Nephila* and *Araneus Diadematus*, which can produce 6–7 kinds of yarn.

Canadian and US scientists have attempted to produce spider silk from goat milk by introducing spider protein-coding genes into goat DNA in order to make it possible to extract the substances necessary for the production of spider silk from the milk. Similar studies are being carried out by European researchers using GM potatoes. The synthetic spider's web would have thousands of applications, from medical (soluble stitches in microsurgeries, artificial tendons) to industrial (light bullet-proof clothes for soldiers and police officers, parachute silk, very strong rope and string for fishing). Unlike silkworms, which for centuries have served as a source of raw material for the textile industry, spiders cannot be bred. They cannot be forced to produce the web and what is even worse, they are cannibals. Attempts to create spider farms have ended with one surviving specimen.

*Vicuna (*Vicugna Vicugna*)*

Vicuna (*Vicugna Vicugna*) is one of two wild South American camelids, along with the wild guanaco. It lives on the high, grassland plateaus of the Andes Mountains, which range from southern Peru to northern Chile and into parts of Bolivia and Argentina. It is a relative of the llama and the alpaca. Vicunas produce small amounts of extremely fine wool, which is very expensive because the animal can be shorn only every three years. The product of the vicuna's fur is very soft and warm. The diameter of elementary fibres is 12–15 μm and the fibre length is in the range of 20–25 mm. Vicuna wool is generally used for textiles. The principal consumer markets for the fabrics are Japan (45%), Italy (35%), the UK (10%) and the USA (10%) [2].

1.2.3 Mineral fibres

The comfort properties of mineral fibres preclude their widespread use in interiors. Glass fibres, ceramic fibres and others can be utilized in cases requiring improved fire resistance. Some of these can resist temperature of greater than 1200 °C (for example, basalt and alumina fibres). Glass fibres were once popular in curtain materials and wall coverings, but they have fallen from favour; they have poor abrasion resistance.

1.3 Interior applications

There are many interior applications for natural fibres as yarns, fabrics and non-wovens – from tapestries, wall coverings, tablecloths, carpets, rugs, floor coverings, upholstery fabrics, bed sheets, window blinds and curtains to towels and mattresses.

Natural fibrous materials are environmentally more user-friendly than man-made fibres. They guarantee optimal comfort of use and generate a positive influence on human physiology. By direct contact with skin and subsequently inhaled air in building interiors, they minimize adverse effects on the people living there. The nerve endings in the body collect stimuli from the skin and subsequently breathing system and finally send information to the brain. This is how the interior environment influences people – creating feelings of either comfort or discomfort.

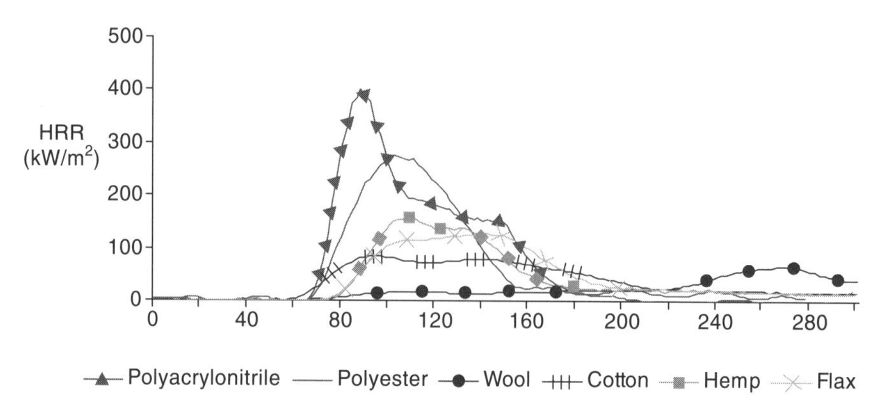

1.7 Heat Rate Release (HRR) (kW/m²) for textiles.
Source: Research work by Institute of Natural Fibres.

Recent research has discovered some new positive features of natural fibres that act as correctors of humidity in interiors, absorb excess UV radiation and are free of some of the nano-particle residues of chemical agents found in man-made fibres such as plasticizers, dyes, bleaching agents and titanium dioxide.

Some natural fibres are bacteriostatic and fungi resistant and can be dyed with eco-friendly natural dyes. However, there are some disadvantages with natural fibres – the main drawback is their non-homogeneity and tendency to biodeteriorate in extreme conditions. On the other hand, the latter disadvantage has its advantage – natural fibres can be easily composted after use and therefore decrease pollution levels caused by waste man-made fibres.

A very important aspect of their use in interiors is their behaviour in cases of fire. Fire Service rules specifies that some textile products, such as carpeting, curtains and upholstery, that are installed in most commercial buildings must comply with flame retardant regulations. Generally, natural fibres have a lower heat release rate compared to man-made fibres and do not drip. The toxic gases emitted during a fire are known and are less dangerous because the rapid oxygen consumption results in shorter burning time (see Fig. 1.7). Smoke emission values are lower for natural fibre fabrics in comparison to man-made fabrics (see Fig. 1.8).

All these properties mean that natural materials are still widely used in interiors and will most likely be so in the future.

1.3.1 Upholstery systems

Natural raw materials in upholstery systems are used mainly as decorative, padding, filling and barrier materials, especially for products of high standards. Natural raw materials are highly valued in various kinds of furni-

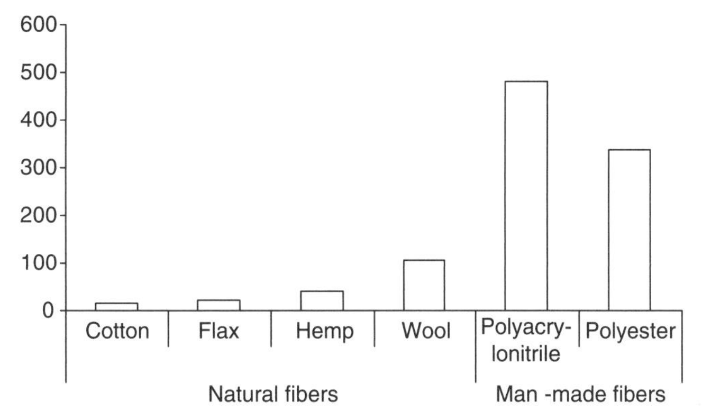

1.8 Smoke emission for textiles – heat flux 35 kW/m^2.
Source: Research work by Institute of Natural Fibres.

Table 1.4 Tensile strength of fibres

Fibre	Value
Polyamide	High
Aramide	↑
Olefin	
Flax	
Polyester	
Silk	
Cotton	
Acrylic	
Modacrylic	
Rayon	
Wool	
Acetate	Low

ture, such as armchairs, mattresses, couches and seats, e.g. for transportation vehicles.

1.3.2 Covering fabrics

When selecting fabrics for a specific application, the decision should involve not only aesthetic aspects but also the fabric structure and composition. In the case of furniture-covering fabrics, the best ones are blends of synthetic fibres (e.g. polyamide, polyester), which are resistant to abrasion and show good tensile strength, and natural fibres (e.g. wool, flax, silk), which are known for their high permeability and moisture absorption capability (advantages missing in the vast majority of synthetic fibres, see Tables 1.4–1.7).

Table 1.5 Light fastness of fibres

Fibre	Value
Acrylic	High ↑
Modacrylic	
Polyester	
Flax	
Cotton	
Rayon	
Acetate	
Olefin	
Polyamide	
Wool	
Silk	Low

Table 1.6 Hygroscopicity of fibres

Fibre	Value
Wool	High ↑
Rayon	
Flax	
Silk	
Cotton	
Polyamide	
Modacrylic	
Acrylic	
Polyester	
Olefin	Low

Table 1.7 Resistance to abrasion of fibres

Fibre	Value
Polyamide	High ↑
Olefin	
Polyester	
Flax	
Acrylic	
Cotton	
Silk	
Wool	
Rayon	
Acetate	Low

New varieties of covering fabrics make use of natural fibres (wool, flax, mercerized cotton), resulting in high-quality soft tweeds and colour-woven fabrics.

The range of natural and cellulose-derived fabrics used for upholstery (especially in public buildings) includes fabrics made of wool, cotton, flax, viscose and their blends, treated with flame-retardant systems that exhibit resistance to washing, e.g. Pyrovatex, Proban, Zirpro [31–33].

Covering fabrics made of wool deserve particular attention as wool has a natural capability for forming intumescent carbonaceous char with excellent thermo-insulating properties. Wool's natural flame retardancy is an important property in, for example, airline upholstery and it is interesting to note that around one tonne of wool is used in furnishing a Boeing airliner [34]. Moreover, in fire conditions, wool blended with natural cellulosic fibres causes a reduction in the extent of the crystalline domains of cellulose and this transformation into amorphous domains reduces the flammability of such blends.

To comply with the fire safety requirements in the presence of small heat sources, usually a specific composition of a fabric, obtained by blending, for example, wool with about 50% cotton or flax, is satisfactory. Moreover, the addition of flax fibres to wool fabrics gives them a pleasant cool feel, which makes them more comfortable to use, protects against moths and prevents static electricity effects.

Leather is also used in the manufacture of luxury products for upholstered furniture and the automotive industry. In these applications, fire behaviour is of extreme importance. Leather, when compared to man-made products used in car and upholstery applications, is safer and is also a healthier choice [62].

1.3.3 Upholstery composites containing natural fibres

Over the years, natural fibres in upholstery composites have been superseded by synthetic fibres, but nowadays the former are again perceived as ecological, renewable and environmentally-friendly materials. This comeback of natural fibres has been accelerated by the possibility of producing materials containing natural fibres with increased resistance to ignition even in the absence of flame retardants [35].

There is a large variety of upholstery composites available on the market. Which to choose is dictated first and foremost by the proposed application. Upholstery composites for furniture designed for standard and luxury apartments will be specified differently from those for furniture designed for public buildings and transportation applications.

From fibres of plant origin (jute, cotton, coir and sisal) to those of animal origin (clipped wool, horsehair and camel hair) – all have found wide areas of application. Although it depends on the usage of the upholstery composite, they are most frequently used in the form of needled fleece and various non-woven fabrics, the properties of which meet extremely specific requirements.

Natural fibres are particularly desirable for furniture designed for reclining/folding, because they determine, to a considerable extent, the comfort achieved in using a particular piece of furniture. For instance, horsehair actively controls airing, humidity and temperature while creating an optimum climate for people who have difficulty falling asleep [36].

For those who prefer sleeping in cool conditions, coir is useful because its structure permits a free flow of air, thus controlling humidity and cooling. It is recommended for the production of firmer upholstery systems and it makes an excellent support for children whose bone tissue is not fully developed [37].

Clipped wool is also highly regarded as a component of composites used in upholstered furniture. The hair is hollow, which causes clipped wool to absorb up to one third of its weight in moisture, and, due to this property, it helps prevent the unpleasant chilliness that results from the vaporization of sweat. Clean wool is also antibacterial, is characterized by high self-purification capabilities and does not absorb any odour. As in the case of cotton, wool is used to cover upholstery composites in the form of needled fleece of different thicknesses and weights.

Polyurethane and latex foams, which have commonly been used for seating, are being replaced by spring mountings covered with non-wovens made of natural materials in furniture designed for comfort. Fulfilling market requirements in the field of upholstered furniture and mattresses, from the point of view of their form, and modern comfort and fire-safety requirements, presents a real challenge to furniture manufacturers. To meet testing requirements, modern fire-resisting components, i.e. non-flammable covering fabrics and flame-retardant filling materials, are usually combined or barrier-layers containing composites are designed [38].

1.3.4 Barrier materials

Another way to protect the inner part of upholstery against fire is the use of barrier materials in the form of non-flammable inter-layers placed between a covering fabric and a standard filling material. Woven and non-woven fabrics, as well as felt made of natural fibres (e.g. cotton, flax and hemp) and their blends with synthetic fibres which have been treated with flame retardants [39], can serve this purpose.

Non-wovens of reduced flammability (intended as barrier materials) based on natural fibres show that fluffy non-wovens made of wool, flax and

Table 1.8 Linen Fire Barrier flammability test non-woven – [Lin FR]300

Flammability test	Unit	Value
Road vehicles ISO 3795-1989(E)	mm/min	0
Oxygen Index (LOI method) ISO 4589-1996	%	28.0 +/– 0.5
Cone calorimeter ISO 5660 (heat flux 35 kW/m^2) Heat release rate (HRR)	kW/m^2	100 +/– 20

hemp are characterized by good flame-retardancy properties. These non-wovens simultaneously play the role of both padding and fire-barrier, thus reducing the vulnerability of a filling material to the development and spread of flame and so enabling the development of difficult-to-ignite non-woven barrier products (see Table 1.8) [40,41]. Recent research [41] reports the development of non-woven barriers based on natural raw materials that provide upholstery systems and mattresses with great comfort of use.

The barrier effects obtained are:

- Flame retardancy
- Antielectrostatic properties
- Biodeterioration resistance (mildew)
- Breathability (air permeability)
- Water sorption
- Resistance to odour.

Bedding and mattresses

Healthy living is the new lifestyle and a good night's sleep is required for this. For example, fifty per cent of Germans sleep badly and one in every three visits to the doctor is related to sleep disorders [47].

The choice of correct bedding has always been considered an important criterion for a good night's sleep. Several studies confirm that wool and flax are the best choices of natural raw materials to guarantee relief from stress and provide good-quality sleep [34, 42]. These fibres provide the best environment for natural all-round comfort and well-being and their unique ability to breathe helps people to keep warm when it is cold and keep cool when it is hot. Wool and linen bedding cools down skin temperatures and provides better moisture management, which is consistent with the natural requirements of the human body. They also offer more comfortable sleep conditions by creating a much lower relative humidity and do not produce adverse physiological effects on the human body. Finally, they give rise to fewer allergic reactions due to wool's inherent bacteria-inhibiting properties.

Research has shown that people sleeping on linen bedding had a high immunoglobulin (α) level and, due to this, they were sound sleepers and their immune systems were stronger [43]. Other studies concerning the health aspects of products made of natural fibres (flax, hemp, wool) have proved that these products do not cause oxidative stress, which is responsible for many diseases including cancer [44, 45].

Products made of the above fibres activate sebaceous glands, which increase the production of squalenes, waxes and triglycerides and consequently create better skin protection against the harmful effects of external conditions [46].

Other combined clinical and laboratory tests [47] show that wool bedding products

- breathe more naturally than man-made synthetic products,
- increase the duration of the most beneficial phase of sleep known as the Rapid Eye Movement stage, where the sleeper is totally relaxed and most dreaming takes place, and
- regulate body temperature – the body gets to a comfortable sleeping temperature more quickly and stays there for longer.

These results all contribute to a better night's sleep.

The Woolmark Company's Australasia general manager, Leah Paff, said 'A bad night's sleep, when you are in and out of sleep, is all related to heart rate. . . . The second your body overheats, your heart rate goes up and down quite erratically, and that's what stops you from being in Rapid Eye Movement sleep stage, or deep sleep' [47].

Natural textile wall-coverings

Natural textile wall coverings are usually laminated to a backing in order to enhance their dimensional stability and to prevent the adhesive from coming through to the surface. These backings are usually acrylic or paper. Their wide range of designs can meet the requirements of everyone: private homes, offices, hotels and restaurants. Products can be treated to produce anti-mould, anti-bacterial, anti-mite, and anti-stain (fire-proof on request) properties and the use of specific dyestuffs gives the fabric high lightfastness properties.

There are many textile wall coverings. One of them is described as: Incotex Clara Lander woven with large vertical flutes (37% cellulose, 59% cotton, 3.5% acrylic, 0.5% viscose or 40% acrylic, 54% cotton, 6% viscose etc.) and an acrylic coat on the back of the material to prevent air from passing through and dust infiltration. As a result, the location is well insulated in terms of protection against noise and against the cold, and energy consumption is reduced [48].

Natural floor coverings

Natural floor coverings are made from renewable resources and are biodegradable, but they have some disadvantages such as: they do not have the stain-resisting qualities of synthetics, they will change colour over time making each floor as unique as the product itself, and they are not UV stable so fading may occur when exposed to harsh sunlight. Due to their absorbent nature, they should not be laid in areas of excessive moisture or humidity because this will cause the carpet to shrink and encourage the growth of mould.

Pure sisal carpets and rugs, carpets in sisal and wool, or in coir, sea grass, jute, bamboo and paper (from conifer trees) are available [49] for applications from doormats to bedroom carpets.

To choose the right floor covering, it is important to know that:

- Coir products are hard-wearing and can be used in a variety of both domestic and light contract locations.
- Jute floor coverings are softer than coir, sisal and sea grass but not as hard-wearing. Placing jute carpet in areas of heavy use such as hallways, staircases, or in areas of direct bright sunlight is not recommended.
- Sea-grass floor coverings are hardwearing and suitable for use in most domestic and light contract areas. They are not recommended for use on staircases.
- Sisal floor coverings are hardwearing and suitable for use in most domestic and light contract areas.
- Wool carpets are softer and warmer than the alternative natural-fibre products. Wool floor coverings are suitable for use in most domestic areas.

From earliest times, wool has been woven into carpets. From the simple, hand-operated looms of traditional craftsmen in exotic locations, to the modern, hi-tech operations of today's multinational carpet manufacturers, woven carpets have earned a prized reputation for durability, longevity, practicality and beauty. Wool is able to absorb moisture well.

Types of woollen carpet [50] include:

- Wilton – the most famous of carpet types derives its name from one of the methods used to make plain woven carpet.
- Axminster – carpets of many colours.
- Tufted carpets – to make tufted carpets, tufts are individually inserted into a woven or non-woven backing using a needling technique.
- Bonded carpets – are produced by inserting tufts into a PVC compound which has been applied to a backing fabric.

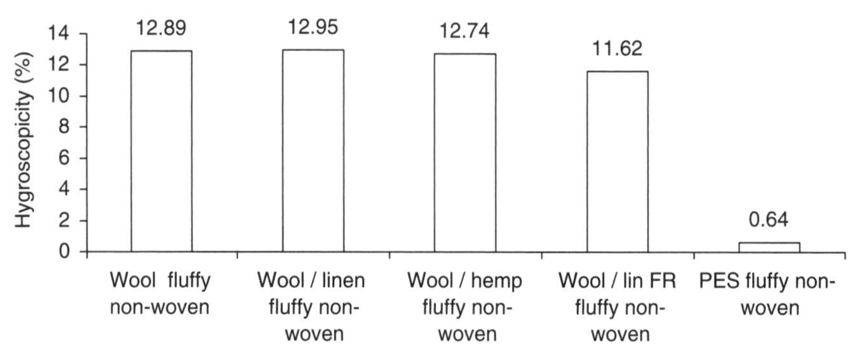

1.9 Determination of hygroscopicity acc. PN-80/P0-4635.

- Knitted carpets – are made using a method in which the backing, stitching and pile yarn are simultaneously combined to form a pile fabric.

For luxury carpets, silk is also very often used.

Insulation materials

For eco-friendly buildings, insulation products made from natural fibres such as wool, flax, and hemp are widely used as thermal and acoustic insulation for walls, ceilings, floors and roofs. These can be found in the form of lightweight boards, non-wovens, mats and foams or as loose material (Fig. 1.9). They have very good thermal and acoustic insulating properties. Flax insulation offers excellent temperature-buffering characteristics, resulting in gradual temperature changes for a comfortable indoor living climate, and very good 'breathing' properties due to its hollow fibre composition, which can have beneficial effects on surrounding constructions as well.

Non-woven insulation materials for houses have been developed from wool and flax as alternatives to insulation made of glass fibre or mineral wool (Isoflax W1) [51].

Other types of flexible thermal insulation are made from hemp and polyolefin fibres. They are particularly economical as they are produced by an environmentally sensitive process (Canaflex). The product is simple to install, and very low in dust.

Advantages of natural fibres, compared with glass fibre, mineral wool and synthetic polymers used for insulation lie in the fact that they are less abrasive, easier to mould and cut, and above all easier to recycle [52].

Interior decorative tapestries

Interior decorative tapestries include:

- woven tapestries
- three-dimensional tapestries
- experimental modern tapestries and
- hand-painted silk textiles

Natural fibres, as a transformation medium, play an important role in different forms of artistic expression. Regardless of the way they are used, whether their natural qualities are emphasized or their colours influence the composition, fibres remain a constructive element in the structure of a tapestry.

Modern art treats fibre as a versatile medium, a flexible material with unlimited possibilities. Perception of a work of art, with its individual character, contributes to a new dialectic discourse between the recipient and the creator.

A study on natural dyeing showed that in nature there are thousands of shades of colours [53].

Original designs based on local traditions, with subdued and soothing colour effects, are well known in the fashion industry nowadays and are frequently an inspiration for modern designers. The experience of using ancient dyes and techniques forges a link with the traditions of the past. As a result, linen, hemp and silk tapestries dyed using natural methods will be created in several ways [54].

In addressing ecological concerns, the primary focus of the art of the creation of tapestries is to show, focus on, and promote the use of natural dyestuffs in textile art. These include natural fibres such as sisal, ramie, abaca, linen, hemp, wool in their natural colours.

Along with traditional tapestries, three-dimensional compositions can be created by changing internal structures.

1.4 Future trends

The overwhelming dependency on petroleum and the increase in legislation on issues related to ecological and environmental impact are two of the main factors determining future developments in the fibre industry. This situation has led to the industry's growing interest in the development of new fibres based on biodegradable and/or recyclable polymers from natural and reusable sources. The main types of fibre with these characteristics are spider silk, alginates, milk casein fibres and polylactic acid fibres [55].

By applying a new technology to the processing of natural plant sugars that are obtained from corn dextrose, the American company Nature-Works has created a proprietary polylactide polymer (PLA). The polymer resins are marketed under the brand name INGEO for fibres. Downstream applications are divided across a full range of apparel and home furnishings, and from carpets to baby wipes [56].

A particularly interesting example of the sophistication of natural polymers is spider dragline silk. Nature has developed a synthetic way of yielding a material with the tenacity of Kevlar® (Du Pont) and the elongation of nylon – an extraordinary product. The genetic engineering route pioneered by Du Pont is probably the most suitable commercial use for it [57].

In recent years, a change in attitude to the problem of environmental protection by consumers as well as producers has been noticeable. The importance of pro-environmental and ethical attitudes is growing fast. In addition, investments in environmental protection have begun to be viewed as aspects of competitiveness in world markets, as well as being investments in the future. The cultivation of cotton has recently been criticized because of its unfavourable effect on the environment – high doses of nitrogen fertilizers, insecticides, and water. Therefore, recently conducted research has focused on developing controlled cultivation of cotton, so-called 'green cotton' or 'organic cotton'. For example, such plantations have been cultivated for ten years in Peru by a Swedish company, Verner Frang AB, as part of a 'White Cotton' programme [58].

Further increases in cotton cultivation are limited by natural barriers such as:

- the limited possibility of increases in efficiency, and
- the accumulation of food crops in the cotton cultivation zones, which are also areas with the highest increases in population

The current main lines of investigation in the development of cotton with improved properties are particularly focused on the attainment of genetically modified cottons that will be softer, wrinkle-free, and endowed with greater dimensional stability.

The best fibrous plant GM-improvement *in statu nascendi* is the synthesis of polyhydroxyalkanates (natural polyester) in living growing plants.

In the challenging world market for home and household textiles, it is essential that new developments are taken up and promoted at an early stage. For example, these days everyone can decorate a bathroom in a harmonious way. New developments are towels made out of bamboo, which has a hollow thread that is particularly good at absorbing moisture, with a yarn with high levels of heat retention that is extremely soft to the touch.

Bamboo decorating items have antibiotic, bacteriostatic, deodorization and ultraviolet-proofing functions. Curtains made from bamboo fibre can absorb ultraviolet radiation in various wavelengths, thus lessening harm to the body [59]. Linen and hemp sauna towels and bath robes have also been suggested.

Additionally, developments in spinning techniques and the application of yarns with a stronger twist have resulted in crepe-like or transparent linen products with increased resistance to shrinkage and creasing.

The wider and wider application of finishing methods that include treatments with resin, liquid ammonia, ultrasound and enzymes in order to soften and reduce the creasing of fabrics can be observed. For instance, in the wool textile industry, as well as in other areas, plasma technology has recently gained much attention because of its enormous potential with respect to surface treatments and process-integrated environmental protection. The plasma technology enables – in combination with a suitable resin – the production of fully washable and shrink-proof wool [60].

More and more fabrics are being produced that incorporate flame-retardant finishes; thanks to technical developments, they look and feel more natural. As well as flame-retardant finishes, additional 'smart' properties of fabrics are considered to be important. Where the commercial and contract sectors are concerned, the new 'high-tech textiles' with the ability to break down harmful or malodorous compounds or with anti-bacterial and anti-static properties are particularly in demand.

The use of nanotechnology provides a great opportunity to change physical and chemical properties on a large scale. The main focus for nanotechnologies is currently on the development of the Lotus-effect on textile structures. The characteristic implied by the expression Lotus-effect is the capacity of a surface to be cleaned completely using only water. This capacity is often called a self-cleaning effect. Research is going on to generate this effect on textiles and to make it permanent. Staining resistance especially is a very important criterion and will lead to a reduction in laundering costs [61].

1.5 Sources of further information and advice

Research institutions

The Textile Institute, Manchester, UK., www.textileinstitute.org
Institute of Natural Fibres, Poznan, Poland, www.inf.poznan.pl
International Wool Textile Organization (IWTO), Brussels, Belgium, www.iwto.org

Tradeshows

HEIMTEXTIL: Frankfurt, Germany, www.heimtextil.de
DOMOTEX: Hannover, Germany, www.domotex.de
NEW YORK HOME TEXTILES SHOW: New York, USA, www.nyhometextiles.com
AMBIENTE: Frankfurt, Germany, www.ambiente.messefrankfurt.com
TEXWORLD: Paris, New York
INTERSTOFF: Frankfurt, Germany, www.interstoff.messefrankfurt.com
DECOSIT: Brussels, Belgium, www.decosit.com

Eco textile tradeshows

Sana: Bologna, Italy, www.sana.it
InNaTex: Wallau, Germany, www.innatex.de
BioFach: Nürnberg, Germany, www.biofach.de
Eco-tradeshow in Schwäbisch Hall, Germany, www.oekomesse-sha.de

Magazines

Journal of Natural Fibres presents new achievements in basic research and the development of multi-purpose applications that further the economical and ecological production of hard fibres, protein fibres, seed, bast, leaf, and cellulosic fibres. An international panel of academics, researchers, and practitioners examines new processing methods and techniques, new trends and economic aspects of processing natural raw materials, sustainable agriculture and eco-friendly techniques that address environmental concerns, the efficient assessment of the life cycle of natural fibre-based products, and the natural reclamation of polluted land.

Textile Month (TM) is a magazine for the international textile manufacturing sector. It has comprehensive news coverage, features and analysis on different spheres of international textile manufacturing.

Ecotextile News is a completely new and unique business-to-business magazine dedicated to the production of sustainable and ethical textiles and apparel.

LDB Interior Textiles is the American world-leading magazine of the home textiles business. Concise articles on specific retailers and vendors as well as business practices of interest to the industry are included in every issue as well as brief news notes of sector activities.

International Textiles Interior – ITBD publishes authoritative magazines on colour trends in the interiors, textiles and body wear industries, and sources new furnishing textiles for contract and residential buyers worldwide.

Associations

World Carpet and Rug Council (WCRC) – The mission is to promote areas of common interest that will facilitate the growth of the carpet industry worldwide and regionally.

British Interior Textile Association (BITA) – is the trade association for the UK interior textiles industry. BITA is regularly consulted by the Government on the industry's views on such issues as changes to the *Furniture & Furnishings (Fire) (Safety) Regulations.* It seeks to promote and safeguard the interests of its members, which range from large, well-known brands to small companies with only a few employees.

The Virtual Interior Textiles Academy (VITA) – is the development and implementation of a virtual multilingual education establishment platform for the European interior textile business.

International Year of Natural Fibres (IYNF) 2009, www.naturalfibres2009.org. The idea is to raise awareness of natural fibres, to promote efficiency and sustainability of natural fibres, and to foster an effective international partnership among the various natural fibre industries.

At the request of FAO, the Food and Agriculture Organization of the United Nations, a declaration was made by the General Assembly of the United Nations on 20 December 2006.

1.6 References

1. BLACKBURN R.S. (2005) *Biodegradable and Sustainable Fibres*, Woodhead Publishing Limited, UK.
2. ROBERT R FRANCK (2000) *Silk, Mohair, Cashmere and other Luxury Fibres*, Woodhead Publishing Limited, UK.
3. FREDERICK T. (2004) *Natural Fibres, Plastics and Composites*, in Wallenberger, Norman Weston Kulwer Academic Publishers, USA.
4. ROBERT R FRANCK (2000) *Bast and Other Plant Fibres*, Woodhead Publishing Limited, UK.
5. AMRO ATEF MOHAMED EL-SAYED (2007) *Studies on improvement of felting and pilling resistance of wool and wool blends via environmentally friendly treatments*, Cairo University, Egypt.
6. XIAOMING TAO (2005) *Smart Fibres, Fabrics and Clothing*, Woodhead Publishing Limited, UK.
7. THE TEXTILE INSTITUTE (1995) *Textile Terms and Definitions*, Tenth Edition, The Textile Institute, UK.
8. HASAN YELMEN (2005) *2400 Years of Turkish Leather*, Derimod, Turkey.
9. *TEXTILES – THE QUARTERLY MAGAZINE OF THE TEXTILE INSTITUTE* (2007) No.2 pages 14–16.
10. *PORADNIK INŻYNIERA – WŁÓKIENNICTWO* (1978) Wydawnictwo Naukowo – Techniczne, Warszawa.

11. DOMAŃSKI W., SURGIEWICZ, J. (2001) *Zagrożenia chemiczne w przemyśle garbarskim*, Bezpieczeństwo pracy nauka i praktyka 4/2001, str. 6–9.
12. DOWGIELEWICZ S. (1954) *Roślinne surowce włókiennicze.*
13. ZIMNIEWSKA M., KOZŁOWSKI R., RAWLUK M. (2004) Natural vs. Man-made Fibres – Physiological Viewpoint, *Journal of Natural Fibres*, volume 1, number 2.
14. AKHILA RAJAN, EMILIA ABRAHAM T. (2007) Coir Fibre – Process and Opportunities: Part 2, *Journal of Natural Fibres*, Volume 4, Number 1.
15. MATHEW, K.U., RANI J. (2006) Isora Fibre: Korphology, Chemical Composition, Surface Modification, Physical, Mechanical and Thermal Properties – A Potential Natural Reinforcement Lovely, *Journal of Natural Fibres*, Volume 3, Number 4.
16. GHITULEASA C., EMILIA VISILEANU E, CIOCOIU M. (2004) Aspect Regarding Romania Angora Mohair Fibres, Characteristics, Application Fields, *Journal of Natural Fibres*, Volume 1, Number 2.
17. ŻYLIŃSKI T. (1950) *Nauka o Włóknie*, W-wa.
18. Saurer Report, *The Fibre Year 2006/2007.*
19. KOZŁOWSKI R., ZIMNIEWSKA M., HUBER J., KRUCIŃSKA I., TORLIŃSKA T. (2002) The Influence of Clothes Made from Natural and Synthetic Fibres on the Activity of the Motor Units in Selected Muscles in the Forearm – *Preliminary Studies, Fibres & Textiles in Eastern Europe*, Vol. 10, No. 4 939, p. 55–59.
20. ZIMNIEWSKA M., KOZŁOWSKI R, Natural or Man-made Fibres – The Question of our Health, *Proceedings of 3rd World Cotton Research Conference*, Cape Town, South Africa.
21. 'Wool Facts – 1986', *Pure New Wool*, september 1986.
22. ORGANIC COTTON MARKET REPORT EXECUTIVE SUMMARY (Spring 2006) – An in Depth Look at a Growing Global Market.
23. KOZŁOWSKI R., ZIMNIEWSKA M, KICIŃSKA-JAKUBOWSKA A. and BOGACZ E. (2007) Electron Scanning Microscope Analysis as a More Effective Tool for Studies on Natural Fibres, *4th Global Workshop of the FAO/ESCORENA European Cooperative Research Network on Flax and other Bast Plants, Innovative Technologies for Comfort*, Romania, 2007.
24. RAFIQ CHAUDHRY M., GUITCHOUNTS A. (2003) *Cotton Facts*, International Cotton Advisory Committee.
25. SIMOPOULOS A.P. (1991) Omega-3 Fatty Acid in Health and Disease and in Growth and Development, *Am J Clin Nutr.* Sep;54(3):438–63. Review. PMID:1908631 [Pubmed – indexed for Medline].
26. THOMPSON, L.U. (2003) *Analysis and Bioavailability of Lignans Flaxseed in Human Nutrition.*
27. THOMPSON, L.U. (2003) *Flaxseed, Lignans, and Cancer Flaxseed in Human Nutrition.*
28. *Nanolignin Modified Linen Fabric as Multifunctional Product* (Book of Abstracts ISBN 978-83-7204-606-3) *IX International Conference on Frontiers of Polymers and Advanced Materials*, 2007.
29. ZIMNIEWSKA M., KOZŁOWSKI R., BATOG J. (2007) Nanolignins as UV Blockers of Textiles, *Biotechnical Functionalisation of Renewable Polymeric Materials.*
30. ZIMNIEWSKA M., KOZLOWSKI R., BATOG J., BISKUPSKA J., KICINSKA A. (2007) Influence of fabric construction, lignin content and other factors on UV blocking, *Textiles for Sustainable Development*, ISBN: 1-60021-559-9, Nova Publishers.

31. HORROCKS A.R. (1989) The Performance of Flame Retardant Finishes – A Critical Assessment, Paper for the *Textile Conference Proceedings*.
32. KOZŁOWSKI R., MIELENIAK B., MUZYCZEK M. (1998) Fire Resistant Composites for Upholstery. *Journal Polymer Degradation and Stability*, October.
33. DAMANT G.H. (1996) Flammability and Material Trends in the Furnishings Industry, *Proceedings FRCA Flame Retardants-101*: Basic Dynamics Past Efforts Create Future Opportunities, Maryland, USA.
34. CHAUDHURI S.K. (2004) Taking Wool into the Future – Through Innovations, *XXI CNTT*, Natal(Brasil).
35. HORROCKS A.R., KANDOLA B.K. (1997) Flame Retardant Cellulose Textiles. *6th European Meeting on Fire Retardancy of Polymeric Materials*, Lille (France).
36. COMMERCIAL INFORMATION PRONATURA. *Naturschlaffibel*, April 1998.
37. FRITZ T.W., HUNSBERGER P.L. (1997) Testing of Mattress Composites in the Cone Calorimeter, *Fire and Materials*, Vol. 21.
38. DAMANT G.H. (1994) Recent United States Developments in Tests and Materials for the Flammability of Furnishings, *Journal of The Textile Institute*, Volume 85 Number 4.
39. KOZLOWSKI R., MIELENIAK B., MUZYCZEK M., KUBACKI A. (2002) Flexible Fire Barriers Based on Natural Nonwoven Textiles, *Fire and Materials*, Vol. 26, 243–246.
40. KOZLOWSKI R., MIELENIAK B., MUZYCZEK M., KUBACKI A. (2002) Flame Retardant Non-wovens on the Basis of Natural Fibres – Production and Application, *Proceedings of the 4th International Wood and Natural Fibre Composites Symposium*. Kassel, Germany, April.
41. KOZLOWSKI R., MIELENIAK B., MUZYCZEK M. (2004) Upholstery Fire Barriers Based on Natural Fibres, *Journal of Natural Fibres*, Vol. 1, No. 1, 85–95.
42. KOZLOWSKI R., MIELENIAK B., MUZYCZEK M., ZIMNIEWSKA M. (2005) Towards the Comfort of the Flexible Upholstery Fire Barriers *FAO/ESCORENA INTERNATIONAL CONFERENCE – 'Textiles for sustainable developments'*, Port Elizabeth, South Africa, 23–27 October.
43. KOZLOWSKI R., MANYS S. (1999) Green Fibres, *Proceedings the 79th World Conference of the Textile Institute*, Chennai, India.
44. BOWERMAN B. (2005) Oxidative Stress and Cancer: A β-Catenin Convergence, *Science*, Vol. 308, 1119–1120.
45. ZIMNIEWSKA M., WITNAMOWSKI H., KOZLOWSKI R., PALUSZAK J. (2002) The Influence of Natural and Synthetic Clothes on Selected Parameters of Oxidative Stress, *The Textile Institute 82nd World Conference*, 23–27 March, Cairo, Egypt.
46. ZIMNIEWSKA M., KOZLOWSKI R., TOKURA H. (2004) Healthful Aspects of Clothing, *XXI Congresso Nacional de Técnicos Têxteis*, Natal, Brazil.
47. *The Study Report from the Woolmark Company*, Heimtextil, Frankfurt, February 2006.
48. Commercial Information: *Incotex Clara Lander*, France.
49. Commercial Information: *Crucial Trading*, England.
50. Commercial Information: *Unique Carpets Ltd*, England.
51. KOZŁOWSKI R., MUZYCZEK M., MIELENIAK B., MANKOWSKI J. (2007) Lightweight, Flexible Insulating Products Made of Natural Raw Materials for Building

Industry, *85th Textile Institut World Conference*, Colombo (SRI LANCA), 1–3 March.

52. SMITH W. (2004) *Report on Regulatory, Commercial and Technical Barriers to the Take up in England of Building Materials Based on Crops*, NNFCC/ DEFRA.
53. SCHMIDT-PRZEWOŹNA K. (1995) Natural Fibres in Unique Textiles, *Fibres & Textiles In Eastern Europe*, NR1(8)/95, p. 62–63.
54. SCHMIDT-PRZEWOŹNA K. (2002) Colour in Polish Modern Tapestry, *Dyes in History and Archaeology, UK*, Archetype Publication, p 125–129.
55. *Vision and Perspective of Consumer Behaviour and Trends in Clothing*, A Global Study executed by Kurt Salmon Associates-IWTO, 2004.
56. COLE M. (2005) Time is Ripe for Bio-Based INGE, *Fibre Journal*, December, 30–32.
57. HEARLE J.W.S. (2007) Protein Fibres 21st Century Vision, *Textiles*, No. 2, 14–18.
58. BERGMAN S. (1999) *Sustainable Development in Textiles Symposium*, Boras, Nov.
59. Processing of Bamboo Fibre in Textile Industries 2007, *Colourage*, April, 2007, 72–74.
60. HOCKER H., THOMAS H. (1999) *International Conference on Advanced Fibre Materials*, Ueda, Oct.
61. STEGMAIER T., DAUNER M., ARNIM V. (2007) and others. Nanotechnologies for Modification and Coating of Fibres and Textiles, *Textiles for Sustainable Development*, Nova Science Publications Inc., New York, 363–371.
62. RYSZARD KOZLOWSKI R., MIELENIAK B., MUZYCZEK M., FIEDOROW R. Flammability and flame retardancy of leather *Leather International*, November, p 24–26.
63. CHANG K., RICHARD P. (2004) Low Dielectric Constant Material from Hollow Fibres and Plant Oil, Wool, *Journal of Natural Fibres*, volume 1, number 2.

2
Synthetic fibres for interior textiles

R. MILAŠIUS and V. JONAITIENĖ,
Kaunas University of Technology, Lithuania

Abstract: Various synthetic fibres are used for interior textiles. The main properties and chemical structures of polyamide, polyester, polypropylene, polyethylene, acrylic and some flame resistant fibres are presented in the chapter, together with their main fields of application. The history and development of the presented fibres is also given, and details of where further information and advice can be obtained.

Key words: synthetic fibres, interior textiles.

2.1 Introduction

All kinds of textile fibres (natural, synthetic, man-made and inorganic) are used in interior textiles. Their fields of application are quite varied, but usually synthetic fibres are used for carpets, curtains, coverlets, pillows and various decorative interior textiles. Many types of materials are used for interior textiles – yarns, woven, knitted, nonwoven fabrics, etc. The six most popular kinds of synthetic fibres used for interior textile manufacturing are polyamide, polyester, polypropylene, polyethylene, acrylic and fibres of the meta-aramid group. Manufacturing of interior textiles also requires non-flammable synthetic fibres, such as meta-aramids and polyamides. The chemical structure and properties of these fibres do not differ from analogous fibres used for clothing or other technical applications. In this chapter we present the main properties and structures of the fibres and focus attention on those properties that are especially important for interior textiles. For further reading, there is the list of articles, websites and books at the end of the chapter, where the various aspects of the structure and properties of these fibres are reviewed in more detail.

2.2 Polyamides

Polyamides (PA) are some of the best-known thermoplastic polymers. In 1935, W. H. Carothers (DuPont, USA) produced the first polyamide (nylon 6.6). The first intended application was as a synthetic replacement for natural silk. Its use started to increase significantly during World War II

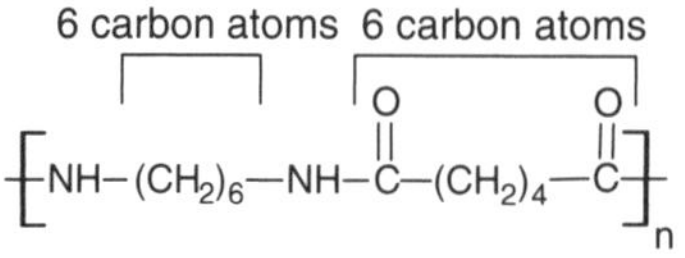

6 carbon atoms 6 carbon atoms

$\left[-NH-(CH_2)_6-NH-\overset{O}{\overset{\|}{C}}-(CH_2)_4-\overset{O}{\overset{\|}{C}}- \right]_n$

(a)

6 carbon atoms 0 carbon atoms

$\left[-\overset{O}{\overset{\|}{C}}-(CH_2)_5-\overset{H}{\overset{|}{N}}- \right]_n$

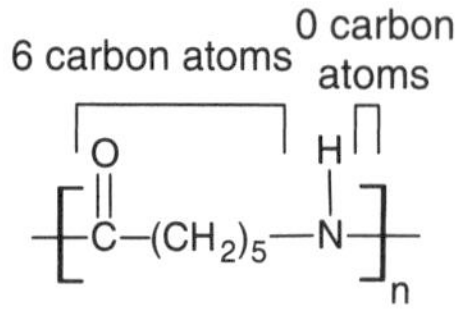

(b)

2.1 Chemical structures of polyamides: (a) polyamide 6.6, (b) polyamide 6.

and, during the rest of the 20th century, it became so popular that all polyamides were called by its first tradename: *Nylon*.

Polyamides are linear macromolecules containing amide groups (–CO–NH–) at regular intervals. The particular type of polyamide depends on the structure of its monomers, which differ by the number of carbon atoms they contain. Thus, there are various kinds of polyamides depending on the number of carbon atoms: PA 4.6, PA 6, PA 6.6, PA 7, PA 11, PA 12, etc. The most commonly used polyamides world-wide are PA 6 (*Kapron* in Russia) and PA 6.6 (see Fig. 2.1).

PA 6 and PA 6.6 are manufactured in different ways: PA 6 is made by polymerizing caprolactam (a cyclic amide derived from a particular amino acid) to polycaprolactam; PA 6.6 is made by condensation polymerization of hexamethylene diamine and adipic acid, through the intermediate PA salt, to poly(hexamethylene adipamide). These two polymers also have different properties: the melting temperature of PA 6 is approximately 215 °C (with a glass transition temperature of 40 °C), while the melting temperature of PA 6.6 is slightly higher, at approximately 250 °C (and with a glass transition temperature of 50 °C). The tensile strength of polyamides is up to approximately 80 cN/tex, whilst the elongation at break is approximately 25%.

Polyamides are light polymers with a density of only 1.14 g/cm^3; they are abrasion-resistant and, due to this property, are very popular in carpet manufacturing. Although not commonly used for other interior textiles, polyamides are the synthetic fibres most used in carpet manufacture, and blended polyamide and wool yarns are often used for this application too. Even small quantities of polyamide (up to 20% of the blend) increase the dynamic properties of woollen yarns and thereby the durability of carpets made from them. Polyamides have a low resistance to sunlight; degradation by sunlight is considerable and they are therefore unfit for curtain manufacture. Polyamides are also not highly resistant to concentrated minerals and organic acids, though they are resistant to alkalis.

Since polyamides are colourless and also easily dyed, their other application is for decorative goods. Polyamides can be dyed topically or in a molten state (solution dying); they can be printed easily and have excellent wear characteristics. However, it is important to note that, due to dye sites on the fibre, polyamides tend to stain very easily, and these dye sites need to be filled to increase polyamide stain resistance. In interior textiles, polyamides are usually used as multifilament yarns, and woven or knitted fabrics are produced from them.

2.3 Polyester

Polyester fibres (PES) are made from polyethyleneterephthalate (PET). PET was invented as a plastic in W. H. Carothers' laboratory (USA) in the 1930s but, after the discovery of nylon, work on polyester was slowed down. In 1939, the British scientists J. R. Whinfield, J. T. Dickson, W. K. Birtwhistle and C. G. Ritchie resumed W. H. Carothers' investigations and, in 1941, created the first polyester fibres which they called *Terylene*. In 1946 the DuPont company (USA) bought all author rights and developed polyester fibres which they called *Dacron*. In 1951 polyester was introduced to the textile and clothing market. After that, other polyester fibres were created: *Kodel* (by Eastman, another USA company), *Trevira* (by the German company, Trevira) and *Lavsan* (in the Soviet Union).

The market for polyester fibres grew until the end of the 1960s but sales drastically fell in the 1970s due to discomfort caused by the fibres, especially in clothing. A new age of polyester use started in the 1990s after the manufacture of microfilament fibres began in 1986. Due to their versatility and easy care, polyester fibres are now one of the most widely used synthetic fibres for clothing and interior textiles.

The macromolecules of polyester contain ester –COO– functional groups, and this is the reason for its name. A benzene ring with two ester groups composes a terephthalate group which, with an ethylene group, forms polyethyleneterephthalate polymer (see Fig. 2.2).

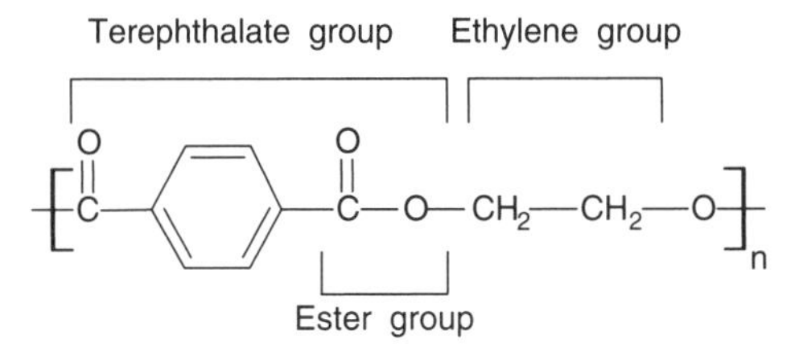

2.2 Chemical structure of PET polyester.

Polyester is not a light polymer – its density is not very low at 1.38 g/cm^3 – and it has a high melting temperature of approximately 250 °C. There are various polyester fibres with different trade names and even with varying properties. In terms of the tensile properties of the fibres, it is possible to distinguish two widely used types of polyester fibre: high strength and medium strength polyester. The breaking force of high strength polyester is approximately 60 cN/tex (elongation at break approximately 10%), while the breaking force of medium strength polyester is approximately 40 cN/tex (elongation at break slightly higher at approximately 15%). High strength polyester is sometimes called technical polyester and is widely used for technical applications, though also for interior textiles, in which case multifilament yarns (and products produced from them) are made. Medium strength polyester is predominantly used for spun and textured yarn production and is very widely used for clothing and interior textiles.

Polyester is widely used for various kinds of textiles as woven, knitted and nonwoven fabrics. It is resistant to acids and alkalis, and is also resistant to sunlight, the last property making it very popular for curtain manufacturing, though it is also used for coverlets, pillows and decorative interior textiles, etc.

A polyester fibre called *Trevira CS* is used for interior textiles that need higher flame resistance. Due to high limiting oxygen index (the LOI of *Trevira CS* is 28%, while that of other polyesters is only 20%), these fibres are sometimes called flame-resistant polyesters and are used in interior textiles for curtains and upholstery manufacture. It is important to note that, due to their thermoplastic nature, the field of their application is sometimes restricted since, like all thermoplastic fibres, flame resistant polyester starts to melt at high temperature.

2.4 Polypropylene

Polypropylene (PP) is a fibre from the polyolefins group. Polyolefin was created in 1966 and there are only two widely used fibres in this group: polypropylene and polyethylene, both polymers being used for interior textiles. Polypropylene's molecular formula is $(C_3H_6)_n$ (see Fig. 2.3).

Polypropylene is a very light polymer with a density of only 0.95 g/cm^3 and has a very low melting temperature of only 160 °C. Its tensile strength

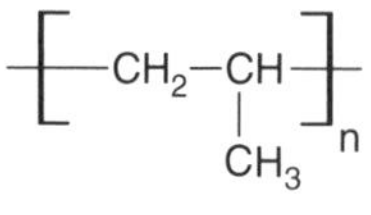

$$\left[-CH_2-\underset{\displaystyle CH_3}{\underset{|}{CH}}- \right]_n$$

2.3 Chemical structure of polypropylene.

is approximately 60 cN/tex, and elongation at break is approximately 20%. Polypropylene is a very abrasion-resistant fibre and, due to this property, it is very popular for upholstery manufacturing. For upholsteries, woven polypropylene is used in various forms: spun yarn, textured yarn or multifilament yarn. Microfilament polypropylene is especially popular for this application, producing fabrics with excellent handling properties. Polypropylene is also used in the carpet industry, especially for bump yarns (for which multifilament yarns are normally used). Polypropylene is highly resistant to acids and alkalis but has very low resistance to sunlight.

2.5 Polyethylene

As mentioned in Section 2.3, polyethylene fibres (PE) are from the polyolefins family and their chemical structure, $(C_2H_4)_n$, (Fig. 2.4), is very close to that of polypropylene.

Like polypropylene, polyethylene is a very light polymer with a density of only 0.95 g/cm^3, and with a similarly very low melting temperature of approximately 150 °C. From the point of view of tensile properties of the fibres, it is possible to distinguish two widely used types of polyethylene fibres: high strength and low strength polyethylene. The tensile strength of low strength polyethylene is approximately 40 cN/tex and the elongation at break is approximately 20%, while the analogous properties of high strength polyethylene are remarkably different.

Manufacture of high strength polyethylene, produced by increasing the crystallization and changing the orientation of molecular chains in the direction of the fibre axis (the degree of orientation), has been theoretically possible since the 1930s. However, it was first created only in 1979 by the Dutch DSM company, who have manufactured fibres under the tradename *Dyneema* since 1986. The strength of the fibres depends on their modification but is up to 370 cN/tex, while elongation at break is only 3.5% (*Dyneema SK 76*). Similar fibres, called *Spectra*, are now also produced by Honeywell, USA; *Spectra 2000* has a strength of 340 cN/tex with an elongation at break of less than 3.0%.

In the field of interior textiles, polyethylene, like polypropylene, is usually used in carpet manufacturing, especially for bump yarns.

$$\left[-\overset{H}{\underset{H}{C}}-\overset{H}{\underset{H}{C}}- \right]_n$$

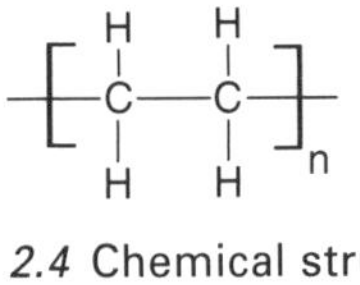

2.4 Chemical structure of polyethylene.

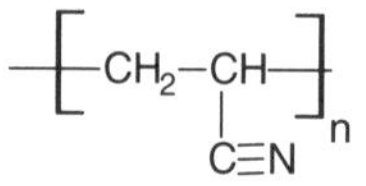

$$-\!\left[CH_2-\underset{\displaystyle C\equiv N}{\underset{|}{CH}}\right]_n\!-$$

2.5 Chemical structure of acrylic (polyacrylonitrile, PAN).

2.6 Acrylic fibres

Acrylic fibres (polyacrylonitrile, PAN) were first made in 1941 by DuPont (USA) and trademarked *Orlon*; just one year later, in 1942, acrylic fibres were also made in Germany. Commercial manufacture however, started only in 1950. Acrylic fibres are produced from acrylonitrile (first made in Germany in 1893). Today there are various trade names for acrylic fibres: *Acrilan*, *Creslan*, *Dolan*, *Filana*, *Pil-Trol*, *Inova*, etc. The polyacrylonitrile linear chain molecule is built from repeating units of CH_2CHCN (see Fig. 2.5).

Acrylic fibres have a high crystalline morphology, a very high melting temperature (for a thermoplastic fibre) of approximately 320 °C, and a glass transition temperature of 90 °C. This is a light polymer with a density of 1.18 g/cm^3. The tensile strength of acrylic is approximately 40 cN/tex and the elongation at break is approximately 20%. A very important property of acrylic fibres is their wool-like handle: they are very soft and warm, and, furthermore, they have a good resistance to light and chemicals. Consequently, acrylic is used for drapes, fur imitation, carpets and other decorative interior textiles. Some modified acrylic fibres have high anti-pilling or flame resistant properties; as is the case with flame resistant polyester, the field of application of these modified acrylic fibres is sometimes restricted due to their thermoplastic nature.

2.7 Flame resistant fibres

Flame resistant fibres are usually used in interior textiles for curtains and upholstery manufacturing, but they are also used in specialist carpet manufacturing. However, it is not just synthetic flame resistant fibres that are used for fire resistant interior textiles: natural wool and cotton fabrics with special flame resistant finishings also have a very large application. These natural fibres, called FR Wool and FR Cotton (FR standing for 'flame resistant' or 'flame retardant'; the terms are interchangeable) are very competitive, especially for interior textiles applications. Of the various synthetic flame resistant fibres, two have the largest application within interior textiles: meta-aramid and polyamide–imide.

Meta-aramid (poly(m-phenyleneisophthalamide)) fibres were first created by DuPont (USA) in the early 1960s and called *Nomex*. Slightly

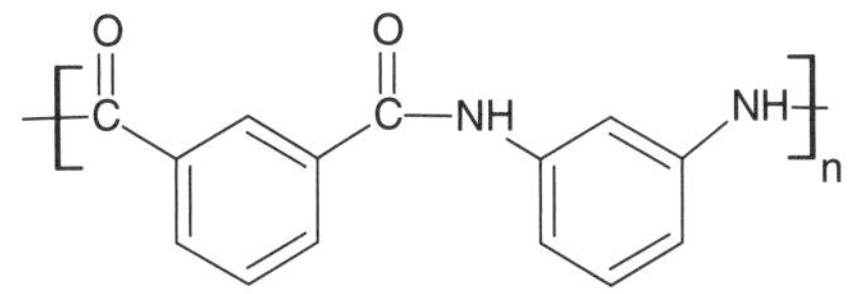

2.6 Chemical structure of a meta-aramid.

later, other meta-aramid fibres were created: *Conex* in Japan (by the Teijin company) and *Fenilon* (in the Soviet Union). Since then, *Nomex* and *Conex* are the most widely used meta-aramid fibres.

Meta-aramid is a fibre from the aromatic polyamide group called aramid fibres. In meta-aramid, the meta-phenylene groups (amide groups) attach to phenyl rings at the 1 and 3 positions (see Fig. 2.6).

Meta-aramid is a high temperature resistant fibre: its decomposition temperature is approximately 500 °C and it is characterized by very long-term stability at 200 °C. It is important to note that meta-aramid is not thermoplastic and does not melt. This positive property expands the field of its application in comparison to flame resistant polyester or flame resistant acrylic. Meta-aramid has a high limiting oxygen index: 30–32, dependent on its various modifications. The density of meta-aramid is 1.38 g/cm^3. It does not have a high tenacity and has a low modulus. However, the initial modulus of meta-aramid is very close to the modulus of high-strength fibres such as para-aramid (*Twaron*, *Kevlar* and etc.).

The breaking strength of meta-aramid is approximately 40 cN/tex and the elongation at break is approximately 20%, but at 20 cN/tex (i.e. 50% of breaking strength) the elongation of fibres is less than 5%. Meta-aramid has a high resistance to acids, alkalis and ageing, even in sunlight, and due to these properties meta-aramid is very popular for interior textiles in chemical laboratories and others rooms where chemicals are used. However, the main application of meta-aramid is for flame retardant/resistant textiles for interiors.

Polyamide-imide is usually attached to the meta-aramids group, but the chemical structure of aromatic polyamide-imide differs slightly from classical meta-aramid in that the fibres have an imide group in their chemical structure. *Kermel* fibres (Rhône Poulenc, France) are the most widely produced of the polyamide-imide group. The properties of polyamide-imide are very close to those of meta-aramid and their application in interior textiles is also the same.

2.8 Sources of further information and advice

AHLUWALIA V.K. and MISHRA A., *Polymer Science. Textbook*, Taylor & Francis, 2008.
ÅSTRÖM B.T., *Manufacturing of Polymer Composites*, Chapman & Hall, 1997.

BLACKBURN R.S., *Biogedradable and sustainable fibres*, Woodhead Publishing Limited, Cambridge, England, 2005.

BOWER D.J., *An Introduction to Polymers Physics*, Cambridge University Press, 2004.

DUBINSKAITĖ K. and MILAŠIUS R., 'Investigation of Dynamic Properties of PA6 and PA6.6 Carpet Pile Yarns, *Materials Science (Medžiagotyra)*, 2005, 3, 288–291.

DUBINSKAITĖ K., VAN LANGENHOVE L. and MILAŠIUS R.,' Influence of Pile Height and Density on the End-Use Properties of Carpet', *Fibres & Textiles in Eastern Europe*, 2008, 3, 47–50.

EBERLE H., HERMELING H., HORNBERGER M. and RING W., *Clothing Technology*, Verlag Europa-Lehermittel, Vokkmer GmbH & Co, 1999.

ESUMI K., *Polymer Interfaces and Emulsions*, Marcel Dekker, Inc, 1999.

FLORY P.J., *Principles of Polymer Chemistry*, Cornell University Press, 1953, (Online http://books.google.com/books?id=CQ0EbEkT5R0C&printsec=frontcover&hl=lt&source=gbs_summary_r&cad=0#PPA218,M1).

HEARLE J.W.S., *High-performance fibres*, Woodhead Publishing Limited, Cambridge, England, 2001.

KIWAŁA M., DAŁEK A. and KRAUZE S., 'Determination of Hydrogen Chloride Concentration during Decomposition and Combustion of Textile Fabrics Used as Interior Decorations', *Fibres & Textiles in Eastern Europe*, 2002, 4, 73–77.

KULSHRESHTKA A.K. and VASILE C., *Handbook of Polymer Blends and Composites*, Volume 1, RAPRA Technology Limited, 2002.

KULSHRESHTKA A.K. and VASILE C., *Handbook of Polymer Blends and Composites*, Volume 2, RAPRA Technology Limited, 2002.

KULSHRESHTKA A.K. and VASILE C., *Handbook of Polymer Blends and Composites*, Volume 3A, RAPRA Technology Limited, 2003.

KULSHRESHTKA A.K. and VASILE C., *Handbook of Polymer Blends and Composites*, Volume 3B, RAPRA Technology Limited, 2003.

KULSHRESHTKA A.K. and VASILE C., *Handbook of Polymer Blends and Composites*, Volume 4A, RAPRA Technology Limited, 2003.

KULSHRESHTKA A.K. and VASILE C., *Handbook of Polymer Blends and Composites*, Volume 4B, RAPRA Technology Limited, 2003.

MATTILA H.R., *Intelligent Textiles and Clothing*, Woodhead Publishing Limited, Cambridge, England, 2005.

MCINTYRE J.E., *Synthetic fibres: nylon, polyester, acrylic, polyolefin*, Woodhead Publishing Limited, Cambridge, England, 2004.

NICHOLSON J.W., *The Chemistry of Polymers*, Royal Society of Chemistry, 2006, (Online http://books.google.com/books?id=FnniJ4MZciQC&hl=lt).

OÑATE E. and KRÕPLIN B., 'Textile Composites and Inflatable Structures', *Computational Methods in Applied Sciences*, Spring, 2005.

RICHARD F. and DALEY S.J., *Organic Chemistry*, Online organic chemistry textbook. (http://www.ochem4free.info).

SPERLING L.H., *Introduction to Physical Polymer Science*, Wiley-Interscience, 2006.

STEIN R.S. and POWERS J., *Topics in Polymer Physics*, Imperial College Press, 2006.

VIGO T.L. and TURBAK A.F., *High-tech Fiberous Materials, Composites, Biomedicals Materials, Protective Clothing, and Geotextiles*, American Chemical Society, Washington, DC, 1991.

ŽEMAITAITIS A., *Physics and Chemistry of Polymers*, Kaunas Technologija, 2001.

3

The use of knitted, woven and nonwoven fabrics in interior textiles

S. J. KADOLPH, Iowa State University, USA

Abstract: Common fabric constructions included in this chapter are knit, lace, woven, and nonwoven. Discussion focuses on these fabric structures and their appropriateness for end use. Differences among fabrics and construction methods are identified in terms of appropriateness for end use, cost, appearance, texture, and performance. Fiberfill, down, leather, suede, and fur are also addressed. The contributions textiles make to interiors are described and some trends in the industry in terms of green or sustainable design are given.

Key words: knit, lace, woven, nonwoven, leather, fabric.

3.1 Introduction

Fabrics made using the techniques included in this chapter are among the most commonly used fabrics in furnishings. In addition, the appropriateness for end use, cost, and aesthetics differ among these construction methods. The construction technique used to produce a specific fabric has an impact on its cost, texture and appearance, and its suitability and performance for a specific end use. Textiles do much to create the mood of an interior through their contributions to colour, texture, pattern or motif, and performance. They soften the contours of the furniture, finish walls and windows, and otherwise enhance the interior. They can be used to absorb sound and make a space more pleasant for work, relaxation, and human interaction.

This chapter will describe knit, woven, and nonwoven fabrics, include some historic information about select fabrics, identify fabric types, describe common furnishing fabrics by name and end use, and suggest performance considerations. The objectives of this chapter are to describe the basic techniques used to produce knit, woven, and nonwoven fabrics; identify commonly used furnishing fabrics made using these textile construction methods; identify fabric performance based on construction technique and fabric type; and explain how these fabrics are used in furnishings. Table 3.1 compares the structures discussed in this chapter.

Table 3.1 Comparison of textile structures

Structure	Structural component	Identifying features	Types	Characteristics
Knit	Yarn	Interlooped structure, one or more yarn sets	Weft knits	Usually one yarn set Yarns move horizontally
			Warp knits	Usually several yarn sets Yarns move vertically with some side-to-side motion
Weave	Yarn	Interlaced yarns, two yarn sets	Plain weave	Balanced: even number of warp and weft Unbalanced: more warp than weft or more weft than warp Basket: at least one yarn set interlaces in groups of 2 or more
			Twill weave	Interlacing pattern creates floats that moves to the right or left by one to create a diagonal wale
			Satin weave	Interlacing pattern where warp and weft yarns float over 4 weft or warp yarns to create a warp or weft faced surface
		Interlaced yarns, two or more yarn sets	Fancy weave	Elaborate fabric structure that creates a pattern as the fabric is woven; types include dobby, jacquard, momie, leno, extra yarn weaves, piqué, and tapestry
			Double cloth weave	Three or more yarn sets that creates double faced fabric, double cloth, and pocket cloth
			Pile weave	Three or more yarn sets create a pile fabric
		Interlaced with three yarn sets	Triaxial weave	Yarns interlace at 60 degree angles
Nonwoven	Fibre	Fibrous web	Based on web creation and bonding processes	Related to fibre length, fibre type, fibre orientation, and process
Felt	Fibre	Wool fiber web	True felt	Compact, weak fabric
Down and feather	Plumage of water fowl	Presence or absence of quills	All down, mixture of down and feathers, all feathers	Loft or fill capacity
Leather	Animal skin or hide	Texture and grain of skin or hide	Leather	Grain, structure, and thickness varies by species and area of skin or hide
			Suede	Brushed or napped surface
			Fur	Leather with hair attached

3.2 Knitted fabrics

Knitting is a fabric construction method where one or more sets of yarns are interlooped. Knitting is a more recent technique compared to weaving. Pieces of knit fabrics dating from approximately 250 BCE were found near the border of ancient Palestine. Knitting was a hand process until 1589, when the English Reverend William Lee invented a flatbed knitting machine that was 10 times faster than hand knitting. Circular-knitting and warp knitting machines were developed approximately 200 years later. Today, complete products such as rugs and table linens can be produced directly on the knitting machine without requiring additional labour to finish edges, as is required with woven textiles.

Knitting is an efficient and versatile fabric construction method. Knitting machines produce fabric at about four times the rates of weaving looms because operating speeds for knitting machines are not limited by machine width, as it is with looms. Nevertheless, the high productivity of knitting is offset by the increased amount and cost of the yarn required to produce a knit fabric. Because of the loop shape, knit fabrics have greater bulk and require more yarn compared to woven fabrics. The porous structure inherent in the loops of knits provides less cover than for woven fabrics. Thus, small stitches (finer gauge) and finer, more-uniform and more expensive yarns are used to produce dense and firm fabrics and prevent the formation of thick-and-thin places in the fabric. The loop structure of knits makes them more prone to snagging and sagging, but finer yarns and more stable structures minimize these problems. The extensibility of knits makes them especially appropriate for fitting around frames of curved furniture.

Commercial knitting machines utilize computer-aided design systems, electronic-patterning mechanisms and quick style change machine modifications. Electronic controls manage the stitch type for each needle, select the yarn to be used in each stitch, and adjust the yarn tension to minimize flaws in the fabric. Electronic knitting machines make changing patterns much simpler and quicker, and increase the production speeds while decreasing fabric flaws. Other innovations include the weft-insertion knitting machine that inserts weft yarn for better crosswise stability and the warp-insertion knitting machine that inserts warp yarns for greater lengthwise stability.

In knitting, needles form a series of interlocking stitch loops from one or more yarn sets. There are two types of knit fabrics: weft knits and warp knits. These names come from the terms for the direction of yarns in woven fabrics. Weft knits are made from yarns that are carried back and forth or around to form a fabric. Thus, the yarns in a weft knit move primarily horizontally. In warp knit fabrics, yarns move vertically through the fabric or parallel to the direction that warp yarns move in a woven fabric. In

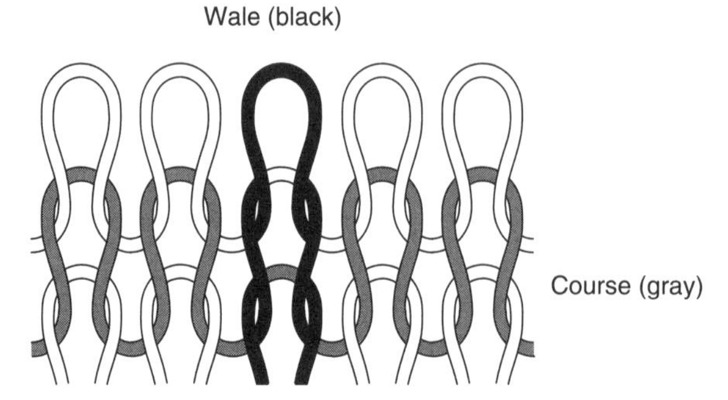

3.1 Weft knit structure with a wale and a course identified. The gray yarn shows the interlooping structure of knits: one yarn forms the wales and the courses.

knitted fabrics, separate sets of yarns do not move in lengthwise and crosswise directions as they do in weaving. Many knit fabrics are made with only one set of yarns. If additional yarn sets are used, these other yarns are used to add visual interest to or improve end use performance for a product. It is fairly easy to unravel most woven fabrics. When a woven fabric is unravelled, both warp and weft yarns are removed. If only warp yarns are removed, a fringe of weft yarns remains in the area where the warp has been removed. If only weft yarns are removed, a fringe of warp yarns remains in that area. When a weft knit fabric is unravelled, a row of loops is removed. However, because the row of loops is the only component in that area, there is no yarn fringe or other portion of the fabric remaining. Most knit fabrics are more difficult to dismantle compared to most woven fabrics. Because of the loop or stitch shape, weft knits can only be unravelled from the end knit last. Because of the warp knit structure, it is very difficult to unravel even a tiny portion of these fabrics since adjacent yarns interlock.

Fabric density for knits is defined by describing the number of stitches and the number of rows of stitches. A wale is a vertical column of stitches in a knit fabric. Because of the loop formation, a wale is formed by both the upward moving and the downward moving sides of the loop. While it appears that two yarns make a wale (see Fig. 3.1), the two sides of a loop actually form only one wale. A course is a horizontal row of stitches. Thus, for a weft knit, a single yarn can be used to form the entire fabric. While this is common in hand knit items, it is not practical for commercially produced knits. In commercial knits, many yarns are used to form a fabric, but, each wale is formed by a single needle. Fabric density is designated as wales by courses per unit of measure, such as 9×7 when using centimetres or 23×18 when using inches. Needles in a knitting machine are used to make the loop or stitch. The names are based on the way the stitches are made.

Gauge or cut indicates stitch fineness and describes the number of needles within a specific distance on the needle bar. Gauge is indicated as needles per cm (npcm) or needles per inch (npi). Fabrics with a higher gauge have finer yarns and generally are considered to be higher quality compared to similar fabrics with a lower gauge. Because finishing may stretch or compress the gauge, a finished fabric may not have the same gauge as the knitting machine on which it was made. A fine gauge machine may have 11 npcm or more, whereas a coarse gauge machine may have 6 npcm or fewer.

Snagging can be a problem with knit fabrics. When a yarn is snagged, it is pulled out and stands away from the knit surface. Shiners, or tight areas, may form on both sides of the course next to the snag because the length of the yarn does not change. Resistance to snagging can be achieved by using finer yarns, smaller stitches, and higher yarn twist in knitting the fabric. If snags are cut away rather than being worked back into the fabric, a run may occur, especially if the fabric is a weft knit or a simple warp knit. A run describes the collapse of a wale, in which each stitch or loop is undone so that only a length of the yarn remains in the area rather than the loop structure. A run occurs in a stepwise fashion when one stitch after another in a wale collapses due to stress on the loops when a yarn is cut. Thus, rather than the knit loop structure of a wale, only straight yarns are visible, marring the surface and creating an obvious visual flaw in the fabric.

3.3 Weft knits

Weft knitting can be done by hand or machine. While weft knits are not commonly used in furnishings, they will be used to describe some basic elements of knit fabrics because weft knits are easier to understand compared to warp knits. In machine knitting, one needle for each wale is set into a machine and the stitch is made in a series of steps. By the end of the final step, one needle has gone through a complete up-and-down motion, and a new stitch has been formed. Needles used for knitting machines are shaped so that the shaft forms a hook at the top, with a latch or other device to close it during certain steps of the knitting sequence. To form a stitch, the needle moves up and the stitch or loop currently on the needle slides down the shaft of the needle. The needle continues to move up until it is in its highest position and the old stitch has moved down to the needle's base. Next, the needle hooks a new yarn and begins its downward stroke. As the needle moves downward, the latch (or other devices) closes the hook's opening so that the old stitch can slide up the needle until it drops off the end of the needle. At the same time, the new yarn caught by the hook of the needle is being pulled into a new stitch when the needle is in its lowest position. Thus, a needle moves up and down in a carefully controlled process. This action is repeated continuously to form a knit fabric. Because each needle in the knitting machine is at a slightly different stage in the process, a wave or undulating motion appears across or around the knitting machine, depending on its type.

Weft knitting is done on two types of machines: flatbed machines, where yarns are carried back and forth, or circular machines, where yarns are carried in an upward spiral like the threads on a screw. In the textile industry, knits are most often classified according to the type of knitting machine used to produce the fabric. While a wide variety of weft knits are possible, most of those fabrics are used for apparel. Weft knit fabrics used in furnishings are limited to a few types. These knit fabrics are usually single weft knits made with a basic knit stitch. Basic jersey, similar to the fabric used for casual shirts, is used to produce lightweight and soft bed sheets. For knit terrycloth sheets, the fabric is made with two sets of yarns: a spun yarn set that forms the pile surface and a heat-sensitive yarn that shrinks when heated. Both yarn sets are knit together to form the fabric that is then heat-set to shrink one yarn set and create the pile loops from the other yarn set. Sliver (slī′ ver) pile knits are constructed using a special machine that knits the fabric ground from yarns and pulls small clumps or tufts of fibre from a sliver to form the pile. A sliver is a loose rope of fibre. The surface fibre pulled from the sliver is held in place by the loop structure of the base knit as well as by a small amount of polymer finish applied to the back or non-pile side of the fabric. These fabrics have a fur-like high or deep pile. The advantage of a sliver knit is that a denser pile can be obtained with sliver than with yarn because the amount of surface fibre is not limited by yarn size or by gauge. When the surface pile combines finer and delustered heat-sensitive fibres with coarser semi-dull, non-heat sensitive manufactured fibres, the fabric can be finished to resemble natural furs. In these fabrics, the heat sensitive fibres shrink with heat and create an under-hair component while the non-heat sensitive fibres create lustrous guard hairs for a more realistic look. The fabric can also be printed to resemble the furs of exotic protected species such as jaguar or leopard, or used in other designs for fun furs. Some fake fur sliver knits are used for casual upholstery fabrics, decorative pillows, and for small rugs such as bath rugs. Fur-like fabrics are lighter weight, more affordable, and more pliable when contrasted with real fur. In addition, these fake-fur fabrics require no special storage conditions as do items made of real fur.

3.4 Warp knits

In approximately 1775, Crane of England invented the tricot (pronounced tree′-ko) machine to make warp knits. Warp knitting is the fastest means

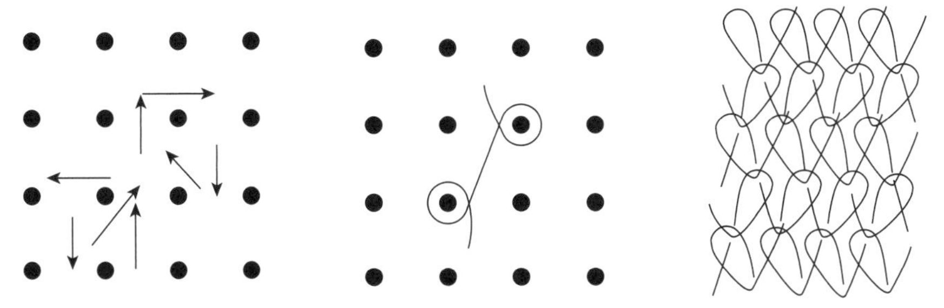

3.2 Warp knit structure: point paper notation for 1 repeat (left); diagram of yarn path (center); warp knit structure (right).

of making fabric from yarns. Warp knits can be considered as combining the best qualities of weft knit and woven fabrics. A warp knit fabric has some degree of elongation with good stiffness, drape, and wrinkle resistance. Warp knit fabrics tend to be less resilient and lighter weight than weft knits. Depending on their structure, warp knits can be stable in either fabric directions or exhibit stretch in only one fabric direction. Warp knits are used in contract-grade carpet, upholstery, drapery and window treatment fabrics, for face fabrics in wall partitions and office panels, and to cover the slats of mini-blinds.

Warp knitting produces a loop structure in a vertical orientation with some slight side-to-side movement of yarns (Fig. 3.2). Warp knitting is a machine process of making flat fabric using one or more sets of yarns. The yarns used in creating the basic fabric structure in a warp knit warp must be very regular. Hence, most often the base structure uses smooth filament or textured filament yarns. The base yarns are fed from warp beams to a row of knitting needles extending across the width of the machine. With rare exceptions, warp knitting machines are the widest type of machine that makes fabric from yarn. Each set of yarns in a warp knitting machine is controlled by metal guides mounted on a guide bar extending across the entire width of the machine. The number of guide bars and yarn sets corresponds with the type of fabric made on the machine. For example, a machine using one set of yarns will have only one warp beam and only one guide bar. This machine will produce the simplest tricot warp knit – a one-bar tricot. A machine with two sets of yarns will have two warp beams and two guide bars. It will produce another type of simple tricot – a two-bar tricot. Each yarn guide on the guide bar controls one yarn and guides it into the hook of one knitting needle. As the number of guide bars increases, greater design flexibility is possible. In warp knitting, a single guide bar directs each operation so that all yarns of one set move identically at the same time. The loops of one course are made as the guide bar rises and moves sideways, moving each yarn around each needle to form loops.

These newly formed loops are dragged down through the loops of the preceding course to complete the formation of one course.

A point-paper notation is used to diagram warp knits. In this diagram, each point in a horizontal row represents a single needle or a wale. Each horizontal row represents a course. Arrows represent the movement of the guide bar and the yarns that form each course. The next row of points represents the next course. This process is done again and again until one complete repeat has been illustrated. To understand the diagram, begin at the bottom row of points and move up the paper from course to course. Figure 3.2 is a diagram of a warp knit stitch (left) and the yarn path (centre) that would be created by that series of guide bar movements. Yarn guided by the front bar most often forms the majority of a warp knit fabric's surface; yarn guided by the back bars provides run resistance, stability, and weight. Most tricot warp knits used in upholstery are at least four-bar warp knits, while warp knit carpets are more complex and require additional bars to create the pile.

Warp knits are classified by the machine used to produce the fabric. Tricot machines use a single set of needles, computer-controlled guide bars, electronic beam control, and computerized take-up. As the fabric is knit, a microcomputer signals the machine to automatically wind the knitted fabric onto a fabric beam to maintain appropriate tension on the yarns. Tricot machines are used to make upholstery, some window treatment fabrics, and some wall partition and office panel fabrics. Raschel machines use one or two sets of needles. Jacquard raschel knitting machines with computer-controlled guide bars produce complex structures used for furnishings. These complex raschel machines are used to make more elaborate upholstery, window treatment fabrics, carpet and some lace.

The lock stitch used in most tricots is shown in Fig. 3.2. The face of the fabric is formed by the short vertical sides of the loops; the back is formed by the long slanted portions of the yarns between or connecting loops. Because of the warp knit structure, tricot does not ravel and lock-knit tricots do not run. One problem with simpler warp knits is that tricots may split or 'zip' between wales if one or more of the yarns is broken and part of the knit structure collapses. While a hole will not appear, the split area is thinner and more susceptible to tearing and abrasion. Tricot is more dimensionally stable than comparable weft knits such as jersey. Tricot has little elasticity in the lengthwise direction and some elasticity in the crosswise direction. Several different weights and styles of tricot are used for upholstery. A second type of stitch, the plain stitch, can also be used to make tricot, but it runs and is seldom used except as the backing fabric for quilted and bonded fabrics for inexpensive upholstery and bedspreads.

Tricot is made with filament yarns used in either smooth or textured form. Brushed or napped tricots have fibres raised from the surface during finishing that create a fabric with the feel and look of velvet or velour. For these fabrics, the knit stitches have long under-laps or horizontal yarn segments between loops. One set of yarns may be carried over three to five wales to form floats on the technical back; the second set of yarns interloops with adjacent yarns. The second set of yarns is usually nylon for its high strength, good durability, and abrasion resistance. The long floats are broken when the fabric is brushed in finishing. The brushed side is used as the fashion side for upholstery and draperies. Tricot-net fabric for inexpensive sheer window treatments can be produced in wide widths by skipping every other needle so only half as much yarn is used and open spaces are created in the fabric.

A double-knit velvet tricot upholstery is made from two layers of fabric that are warp knit face-to-face with a pile yarn connecting the two base layers. The frequency with which the pile yarn connects the two layers determines the pile density. The layers are separated when the pile is cut. Pile height is half the distance between the two layers. Varying the size, type, lustre, colour, or texture of the pile can be used to create a design as the fabric is being knit. For example, knitting two slightly different textured pile yarns in small geometric shapes will create a subtle pattern when the pile is brushed. In another option, fibres that are modified to react to different dyes can be used so that, when dyed, bi-coloured or more complex colour patterns can develop.

The raschel-warp knitting machine has one or two needle beds with as many as 78 guide bars for very complex fabrics. The fabric comes off the knitting frame in a vertical orientation, instead of in a horizontal orientation as with the tricot machine. Raschel machines are used to knit a wide variety of fabrics, from gossamer-sheer window treatment to very heavy carpets. Raschel fabrics have rows of chain-like loops called pillars, with laid-in yarns in various lapping configurations (Fig. 3.3). Some window-treatment fabrics are knit on a raschel warp knitting machine.

Carpets have been knitted since the early 1950s. Since production is faster, warp knitted carpets are cheaper to make than woven carpets. Knitted carpets have two- or three-ply warp yarns for lengthwise stability, laid-in crosswise yarns for body and crosswise stability, and pile yarns. Knitted carpets can be identified by looking for chains of stitches on the underside. They seldom have a secondary backing since the structure locks the yarns in place. Warp knit carpets do not require the protection and stability of a backing as is used in tufted carpets. Warp knit carpets are usually designed for commercial or contract uses.

Warp knits also provide an option of inserting yarns during knitting to create a wide range of fabrics. Insertion of yarns in the warp knit structure is a simple concept. Yarns are laid in the bottom of the loops of a course during the knitting process. The laid-in or inserted yarns do not form any

3.3 Raschel warp knit diagram showing pillars (white yarns).

stitches. These laid-in yarns provide directional stability and can be in any direction or angle. Fabric characteristics can be engineered for desired properties. Yarns not appropriate for knitting can be used, such as extremely coarse, fine, or irregular yarns and yarns of fibres such as carbon and glass that have low flexibility. Insertion fabrics are used in window treatments and wall coverings. Weft insertion is done by a warp knitting machine with a weft-laying attachment. An attachment steadily feeds a single weft yarn back and forth into the needle zone or stitching area of the machine, forming a firm selvage on each side. More complex attachments can supply sheets of weft yarns to conveyors that also travel back and forth across the machine. A cutting device trims weft yarn ends along the selvages.

There are several kinds of inserted warp knits: weft insertion with additional yarns in the horizontal direction, warp insertion with additional yarns in the vertical direction, warp and weft insertion with additional yarns in both horizontal and vertical directions, and diagonal insertions with additional yarns at one or more angles throughout the fabric. Weft insertion fabrics combine good strength, comfort, cover without bulk, and weight. These weft insertion warp knits have the increased crosswise stability of woven fabrics but retain the pleasing hand and comfort of knits. In warp insertion warp knits, the inserted yarn is caught in a vertical chain of stitching. The insertion of a warp or vertical yarn in a knit structure gives the fabric the vertical stability of woven cloth while retaining the horizontal stretch of knit fabric. These fabrics are used for curtains, draperies, wall panels, table linens, and wall coverings. Fabrics with both warp- and weft-insertion have characteristics very similar to woven fabrics. These fabrics can be much less expensive than woven fabrics and are available in wide widths. They are frequently used as window-treatment fabrics. Do not confuse inserted warp knit fabrics with other types of fabrics. Careful examination of the fabric will show that a warp knit fabric includes fine filament yarns knitting around the inserted yarns, not through them.

A final fabric construction option using warp knitting machines is the knit-through fabric. In these fabrics, a fibreweb structure forms the substrate and a warp knitting machine knits fine warp yarns through the fiberweb to create a more opaque, yet inexpensive, fabric. With the knit-through structure, as with insertion warp knits, additional yarn sets can be laid into a course as it is knit to create more interest in the fabric. *Arachne*® and *Maliwatt*® are trade names for fabrics made by warp knitting yarns through a fiberweb structure. *Malimo*® uses warp, weft, or both warp and weft laid-in yarns. These knit-through fabrics are produced at high speeds and are used for upholstery, blankets, tablecloths, wall partitions, office panels, and window treatments.

3.5 Lace

Lace is another basic fabric made from yarns, but several different fabrication methods can be used to produce lace. Most commercial lace is made using a raschel knitting machine. In lace, yarns are twisted around each other to create open areas. Lace is an openwork fabric with complex patterns or figures. Lace is classified according to the way it is made and the way it appears. Remember, warp knitting is only one option for constructing lace. Lace is often used as a trim or accessory in furnishings. Lace and curtain nets can be made at much higher speeds and lower costs on a raschel machine than by other methods. Raschel machines can be programmed to create complex patterned lace, imitating many varieties of expensive hand-made lace as well as Leavers lace, another machine-made lace. However, because raschel laces are produced at much higher speeds, they are less expensive and have largely replaced Leavers lace. Filament yarns are commonly used to make coarser laces that are used as tablecloths, draperies, and casement fabrics. Window treatment nets of polyester with square, diamond, or hexagonal meshes are made on tricot machines. Light, delicate, and elaborate laces can be made quickly and inexpensively using either type of warp knitting machine. Cordonnet, or re-embroidered, lace has a yarn or cord outlining the principle or major elements of the lace design.

3.6 Weaving fabrics

Weaving is one of the oldest and most widely used methods of making fabric. With a few minor exceptions, woven fabrics are made using at least

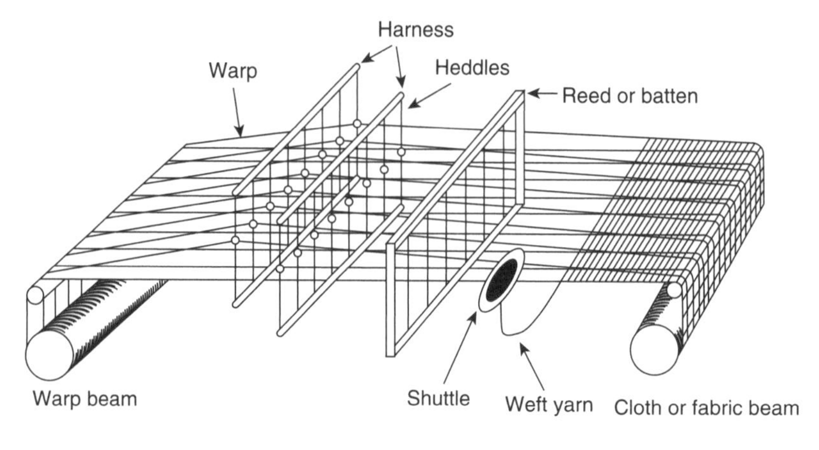

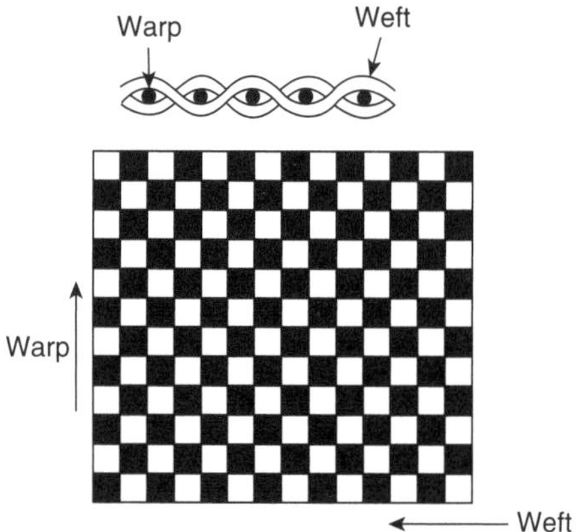

3.4 Diagram of a loom (top) and two methods of diagramming a plain weave fabric: cross-section (center) and checkerboard (bottom).

two sets of yarns interlaced at right angles: warp yarns and weft yarns (see Fig. 3.4). All weaves known today have been made for thousands of years. Woven fabrics are produced using a loom that controls how the warp and weft yarns interlace. The loom has undergone significant modifications, but the basic principles and operations remain the same. Warp yarns are threaded under tension through the loom; weft yarns are inserted and pushed into place to make the fabric. The lengthwise yarns are warp yarns or ends, and the crosswise yarns are weft yarns or picks.

The modern loom (Fig. 3.4) has the warp yarn held under tension between two beams: a warp beam and a take-up beam. Warp yarns are wound onto the warp beam and threaded through the eye of a small wire heddle which is grouped with other heddles in two or more harnesses or frames. The woven structure is controlled by raising the harnesses. There are as many heddles as there are warp yarns in the fabric, but most commercial looms have only four to six harnesses. Each warp yarn passes through the eye of only one heddle. Using this arrangement of heddles and harnesses and controls to raise the harnesses, the warp yarns are separated into groups. One or more groups are raised while the other groups remain at their original position. By raising some of the warp yarns, a shed or opening is formed through which the weft yarn is inserted. One of several devices deposits or inserts a weft yarn in the shed. A reed, or batten, is a set of wires in a frame. The warp yarns are threaded through spaces or dents in the reed that also helps control warp yarn density in the woven fabric. The reed swings forward after each weft yarn is inserted to force it into place close to the previous weft yarn. Reeds are available with a variety of dent sizes. The reed size selected for a fabric is related to the desired density of warp yarns in the finished fabric and the size of the warp yarns. Woven fabric is wound onto the take-up, fabric, or cloth beam as it is woven. The most frequent type of commercial loom is a four-harness loom. This highly versatile loom is used to produce most basic woven fabrics. These fabrics comprise the greatest percentage of woven fabrics currently on the market. Additional harnesses or other devices that control the position of the warp yarns are needed to produce more intricate designs. Fabrics that require more than six harnesses are made on looms that use other devices to control the warp yarns and fabric structure.

Woven fabrics tend to be firmer and more rigid, with little stretch and controlled drape compared to knits, braids, and laces. Adjacent sides can be ravelled. When one set of yarns are removed, a fringe of the other set of yarns remains. When several weft yarns are removed, a fringe of warp yarns remains. If several warp yarns are removed, a fringe of weft yarns remains. Woven fabrics vary in the yarn interlacing pattern, the density or number of yarns per inch, and the ratio of warp to weft yarns. All these factors influence the cost, aesthetics, and performance of a fabric. Several terms relating to woven fabric structure must be defined before further discussion can occur. An interlacing is the point at which a yarn changes its position from the face side of the fabric to the reverse or back side of the fabric. When a yarn crosses over more than one yarn at a time, floats are formed and the fabric has fewer interlacings. Warp yarns are the lengthwise yarns that are threaded through the loom and which are used to control the interlacing pattern and the woven structure. Weft yarns are the crosswise yarns that are inserted to interlace with the warp yarns during weaving. These terms are used to describe almost every weave used to produce woven fabrics: interlacing, float, warp yarn, and weft yarn.

Because the fabric structure is based on yarns interlacing with each other, woven fabrics have grain. Grain describes the angle of warp yarns relative to weft yarns in the fabric. On-grain fabrics have warp yarns parallel to each other and perpendicular to the weft yarns. These weft yarns should form a straight line across the fabric. Fabrics are almost always woven on-grain. Handling, finishing, or stress due to yarn twist or weave structure may distort fabrics forcing them to be off-grain. Off-grain fabrics do not drape properly or hang evenly, and printed designs are not straight. Off-grain fabrics do not work well for draperies, wall coverings, or upholstery, especially when patterns are required to match at seams and fabrics are required to hang evenly and straight without twisting.

Warp and weft yarns often differ in their structure or fibre type. Because of these variations, some fabrics differ in their performance between warp and weft directions. Because the warp must withstand loom tension and abrasion from weaving, the warp yarns are usually finer, stronger, more uniform, and more tightly twisted. Weft yarns are more often fancy or special-function yarns such as high-twist crepe yarns, low-twist napping yarns, or bouclé yarns. Differentiating between warp and weft is possible by carefully examining the fabric and the lengthwise and crosswise yarns. The selvedge is parallel to the warp direction for all woven fabrics. Most fabrics have lower elongation in the warp direction. Warp yarn count tends to be higher than weft yarn count. If yarn-dyed stripes are in one direction only, they are most often found in the warp direction.

Fabric names are based on many factors: fabric structure, yarn type, fabric count, balance, fabric weight, and finishes. Fabric count, count, or fabric density is the number of warp and weft yarns per square centimetre of gray goods (fabric as it comes from the loom). Count may increase due to shrinkage during dyeing and finishing. Count is written two ways: with the warp number first (32 × 30, read as 32 by 30) or as a total of the warp plus the weft (62 for this example). Count contributes to the quality of the fabric; higher count fabrics are better quality.

Balance describes the ratio of warp yarns to weft yarns in a fabric. A balanced fabric has approximately one warp yarn for every weft yarn, or a ratio of 1:1 (read as one to one). An example of a balanced fabric is 60 × 60 print cloth. An unbalanced fabric has many more of one set of yarns than the other and may have ratios of 2:1, 3:1, or more. Unbalanced fabrics may exhibit wear more quickly since one set of yarns is exposed to a higher percentage of the abrasive forces. For such unbalanced fabrics, slits may occur as the wear becomes excessive and the surface yarns are destroyed.

Fabric weight (or more correctly, fabric mass) describes the weight of fabric for a given area or a given length. Both length and area weight values are used in the textile industry. For example, grams per meter or yards per pound may be used in the trade to identify current prices for basic fabrics, but fabric width must be identified in this system. Another system uses weight in grams per square meter (g/m^2) or ounces per square yard (oz/yd^2). Fabric weight is used to identify fabric appropriateness for end use and in naming fabrics. Lightweight fabrics are soft with good drape and comfort and are used for sheets, curtains, sheer draperies, and as substrates or backing fabrics for wall coverings, upholstery, and bonded and quilted fabrics. Medium-weight fabrics are used for warmer sheets, draperies, upholstery, wall coverings, table linens, and for one or more of the layers in quilted and bonded fabrics. Heavyweight fabrics are durable and stiff and are used for more durable upholstery, draperies, and bedspreads.

3.7 Structure of woven fabrics

Woven fabric construction can be represented in several ways. Understanding these representations will help in identifying weaves. A basic plain weave will be used to explain each representation of fabric structure. Figure 3.4 shows two ways of diagramming a woven fabric. The drawing in the centre is a cross-sectional view of a fabric cut parallel to a weft yarn, with the warp yarn ends appearing as black circles. A weft yarn goes over the first warp yarn and under the second warp yarn, repeating the over-and-under pattern all the way across the fabric. The second weft yarn in a plain weave goes under the first warp yarn and over the second warp yarn, repeating that pattern across the fabric. In basic woven fabrics, the interlacing pattern relates to the number of warp yarns the weft yarn passes over or passes under, and a pattern can be readily seen using a cross-section diagram.

In the checkerboard pattern in Fig. 3.4, each square represents one yarn on the surface of the fabric. Squares representing a warp yarn on the surface are black and squares representing a weft yarn on the surface are white. Starting at the upper left-hand corner of the checkerboard and moving across the row, a weft yarn is on the surface followed by a warp yarn on the surface, and so on. The second row reverses the black and white squares because it represents the interlacing pattern of the second weft yarn. All woven fabrics can be diagrammed using this technique, but the pattern will differ with the weave being diagrammed. The simple weaves have the same interlacing pattern throughout the fabric.

3.8 Plain woven fabrics

The plain weave is the simplest weave and is formed by yarns at right angles passing alternately over and under each other. Each warp yarn interlaces with each weft yarn to form the maximum number of interlacings (Fig. 3.4). The technical face and back of a plain weave are identical. However, because of this simple and basic surface, plain wovens serve as a good ground for printing or finishing that creates a technical face. Interesting effects can be created by combining filament or staple yarns, yarns of different sizes, structures, or twist amounts, or by applying surface finishes.

3.8.1 Balanced plain woven fabrics

The simplest plain woven fabric is a balanced plain weave constructed using warp and weft yarns that are the same size and that show in equal amounts on the surface. Plain woven fabrics can be made in a variety of weights, from light to very heavy.

Lightweight sheer fabrics are very thin, weigh very little, and are transparent or semitransparent. Most fabrics in this category are used for lightweight curtains and draperies. Gauze or theatrical gauze is a sheer stiff fabric due to the use of starch or other sizing agents. In ninon, pairs of warp yarns (usually of filament fibres) are spaced close to each other with a slightly larger space between warp yarn pairs. Organdie is an all-cotton fabric that has been treated with a weak acid to produce a slightly stiffer hand and drape. All polyester organza is a filament yarn fabric similar in weight and sheerness to organdie. Both fabrics are used in sheer curtains. Voile is a slightly less transparent fabric made with highly twisted spun yarns used for sheer curtains.

Medium-weight fabrics are the most common of the woven fabrics. These fabrics have medium-sized yarns and a medium count. They may be finished in different ways or woven from dyed yarns. Medium-weight fabrics are used for wall- and window-treatment fabrics, bed and table linens, and some upholstery fabrics. Percale is a smooth, slightly crisp, print or single colour fabric made of combed yarns. In percale sheets, fabric densities of 400 and higher are available. In sheeting, counts are usually reported as totals of the warp yarns plus the weft yarns. Similar fabrics include calico with a small, quaint, printed design; chintz with a print design; and cretonne with a large-scale floral design and slightly coarser yarns. Glazed chintz for upholstery is usually a heavier weight fabric than that used for draperies. Chintz is also used for slipcovers for upholstered furniture. Single colour fabrics with a highly glazed surface are called polished cotton. Printed fabrics with a surface glaze are called glazed chintz. Polished cotton and glazed chintz are used for matching curtains and bedspreads, upholstery, slipcovers, draperies, and wall coverings. Any plain-woven, balanced fabric of carded yarns with total yarn number counts of 110, 130, or 140 is known as muslin. These medium weight plain woven fabrics are used primarily for curtains, but some are also used for draperies depending on the print motif and fabric weight. Medium weight crash composed of linen with its slightly irregular yarn structure is made into lint-free towelling for use in drying dishes and glassware.

Napped fabrics may be of either medium weight or heavyweight. Flannelette is lightly napped on one side. It is available in several weights and is used for sheets and blankets and may be solid colour, yarn-dyed, or printed. Gingham is a yarn-dyed fabric in checks, plaids, or solids. When filament yarns are used, these fabrics are given a crisp finish and called taffeta. When they are made of wool, these yarn dyed balanced plain woven fabrics are called wool check, wool plaid, and shepherd's check. Madras, or Indian madras, is usually all cotton, and has a lower count than gingham. Pongee is a filament-yarn, medium-weight fabric with a fine warp of uniformly regular yarns combined with weft yarns that are irregular in size or that contain slubs. Pongee was originally made of silk with duppion (slub) weft yarns, but is now made of a variety of fibres. Honan is similar to pongee, but it has slub yarns in both the warp and the weft.

Plain-woven fabrics with crepe yarns in either warp or weft or in both warp and weft are true crepe fabrics. They can be any weight, but are most often medium weight or heavyweight. Because of their interesting texture and lively drape, which develops because of the extremely high twist in the crepe yarns, these fabrics are frequently used by designers for furnishings. True crepes are common in upholstery, draperies, wall coverings, and some table linens.

Heavyweight fabrics drape well and are more durable. Homespun has slightly irregular yarns, a lower count, and a hand-woven look. Crash has thick-and-thin yarns and is often linen or a manufactured fibre or fibre blend that resembles linen. Burlap or hessian, usually made of jute, is used in wall coverings and has thick-and-thin yarns. Osnaburg is a variable-weight fabric made with somewhat coarse spun yarns that may include pieces of cotton leaf and stem. It is used as drapery lining, upholstery support fabric, or as substrate for tufted upholstery fabric. It may also be printed or dyed and used in upholstery and draperies. Flannel upholstery may be constructed as a plain or twill weave. Tweed is made with novelty yarns with nubs of different colours. Expensive upholstery fabrics may carry certified registered trademarks designated by ™ to indicate that specific requirements and quality measures have been met. Harris Tweed™ is hand-woven in the Outer Hebrides Islands. Donegal Tweed™ is hand-woven in Donegal County, Ireland. Heavyweight balanced-plain-woven fabrics are used in wall coverings, upholstery, and draperies.

3.8.2 Unbalanced plain woven fabrics

Unbalanced plain woven fabrics have significantly more yarns in one direction than the other. Increasing the number of warp yarns until the count is about twice that of the weft yarns creates a crosswise or weft ridge where the warp yarns completely cover the weft yarns. This type of fabric is warp faced because the warp yarns form the fabric's surface. Warp-faced fabrics are more durable than weft-faced fabrics because the warp yarns are more tightly twisted and more uniform compared to weft yarns. Ribs in fabric are produced by pushing more yarns into the same area or by changing yarn size. Small ridges are formed when the warp and weft yarns are the same size; larger ridges are formed when the weft yarns are larger than the warp. Slippage can be a problem in filament yarn ribbed fabrics, especially those with lower counts. Slippage occurs when one set of yarns is pushed to one side exposing the yarns that are normally covered. This may occur at points of wear and tension, most often at seams. With unbalanced plain woven fabrics, wear occurs on the surface of the ribs. The surface yarns wear out first, and splits occur in the fabric. Ribbed fabrics with fine ribs are softer and more drapeable than comparable balanced fabrics. Fabrics with large ribs have more body and are stiffer.

A few sheer rib fabrics are used in curtains and sheer draperies. Batiste has a slight crosswise or weft rib because of the slightly larger weft yarns used to weave the fabric. Batiste may be used as the ground fabric for tambour-embroidered curtain panels. Medium weight fabrics include taffeta, a fine-ribbed, filament-yarn fabric with a crisp hand. Iridescent taffeta has warp and weft yarns of different colours. Moiré taffeta has a water-mark or embossed design. Shantung has an irregular ribbed surface produced by long, irregular slubs in weft yarns. Medium weight unbalanced plain woven fabrics are used for upholstery that will be exposed to lower amounts of wear, draperies, wall coverings, bedspreads, and table linens.

Heavyweight ribbed fabrics include poplin with heavy and pronounced ribs because of larger weft yarns. Faille (pronounced file) has a fine, subtle rib with filament-warp yarns and spun weft yarns. Rep (or repp) is a heavy, coarse fabric with a pronounced rib. Bengaline has a slightly more pronounced rib that may include double warp yarns to emphasize the rib. Ottoman has alternating adjacent large and small ribs, created by using weft yarns of different sizes or by using different numbers of weft yarns in adjacent ribs. Bedford cord is used in upholstery and bedspreads and has spun warp yarns larger than the weft yarns. Other ways of producing bedford cord will be discussed later in this chapter in the section related to fancy woven fabrics. Heavyweight unbalanced plain woven fabrics are used for heavy-duty upholstery, draperies, and bedspreads.

3.8.3 Basket woven fabrics

In the basket weave, the interlacing pattern is similar to a plain weave, but two or more yarns are grouped to follow the same parallel path (Fig. 3.5). These yarns are not twisted around each other as would occur with a ply

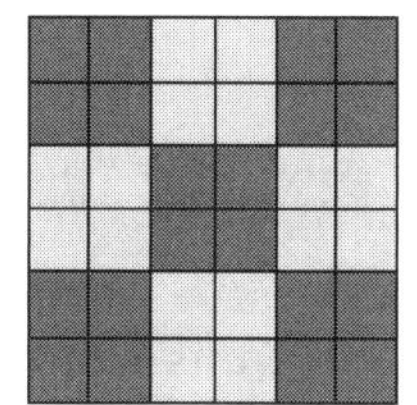

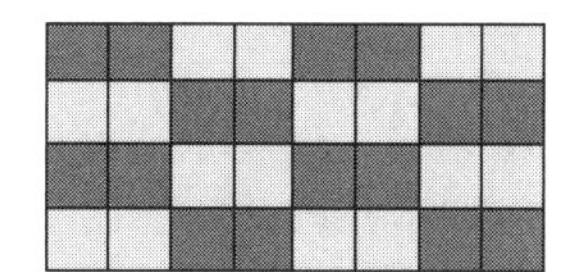

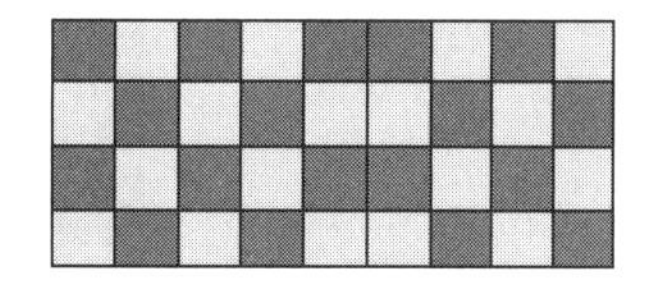

3.5 Basket weave: 2 × 2 full basket (left), 2 × 1 half basket (center), and basket feature (right).

yarn, but rather, are inserted in side-by-side heddles in the same harness for warp yarns or are inserted side-by-side in the same shed for weft yarns. A full basket has both warp and weft grouped and following the same parallel path. In a half basket, only one yarn set (either warp yarns or weft yarns) would be grouped. A fabric with a basket feature would have an occasional grouping of some warp or weft yarns. The most common basket weaves are 2 × 2 or 4 × 4, but such variations as 2 × 1 and 2 × 3 are also common. Basket-weave fabrics are more flexible than other plain-woven fabrics. The fabrics have a flatter look. Because the yarns float for a greater distance on the fabric's surface, they are more prone to snagging and distortion due to abrasion or weight. Dimity is a sheer unbalanced fabric with a basket feature, used for window treatments. It has heavy warp cords at intervals across the fabric. The cords may be formed by yarns larger than those used elsewhere in the fabric, or by grouping yarns together in the cord area.

Most basket weaves are heavyweight fabrics: sailcloth, duck, or canvas. Sailcloth is the lightest in weight and made of single yarns. Duck and canvas are made with single or ply yarns. Different types of duck and canvas relate to which yarns (warp or weft) are plied and how many plies are used to make the ply yarn. Duck is coarser. Canvas is smoother, more compact, and the heaviest of the three. Sailcloth is the heavyweight basket weave fabric most often used in interior furnishings. Usually woven in 2 × 1 or 3 × 2 basket weaves, canvas and duck are used for slipcovers and house and store awnings. Hopsacking is a coarse open basket-weave fabric of spun yarns used for upholstery and wall coverings. Monk's cloth, friar's cloth, druid's cloth, and mission cloth are used primarily in window and wall coverings and casual upholstery. They are usually off-white in colour and available in square counts: 2 × 2, 3 × 3, 4 × 4, or 6 × 6, meaning that the same numbers of yarns are grouped in both the warp and weft directions. Usually 6 × 6 is the maximum number for a square count. Numbers higher than that are too soft and too easily snagged for most end uses.

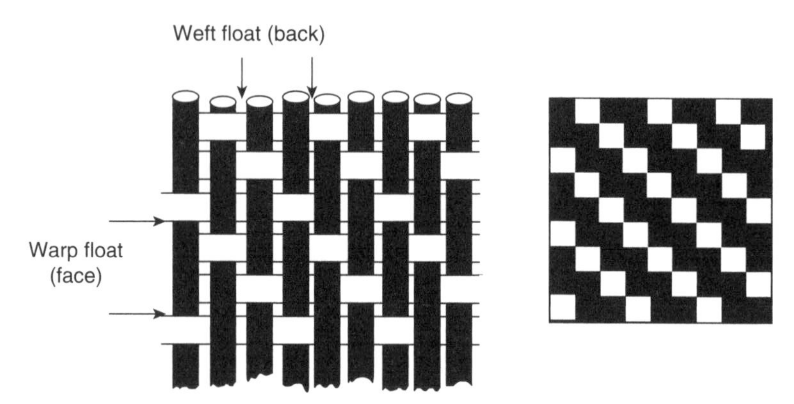

3.6 A twill weave interlacing, with arrows showing the length of warp and filling floats (left) and checkerboard pattern (right).

3.9 Twill woven fabrics

A twill weave has each warp or weft yarn floating across two or more weft or warp yarns, with a progression of interlacings by one to the right or left that forms a distinct diagonal line, or wale. A twill weave is the second basic weave that can be made on the simple loom. It is possible to create fairly elaborate broken or fancy twill weaves, but the majority of this section will focus on basic twill fabrics.

Twill weave structure is often designated by a fraction in which the top number indicates the number of warp yarns on the surface and the bottom number indicates the number of weft yarns on the surface. A basic ${}^{2}\!-\!{}_{1}$ twill is shown in Fig. 3.6. The distance between the arrows demonstrates the length of the float. Since the floats on the surface of the fabric are warp yarns, this fabric would be classified as a warp-faced twill. Twills are also described by the direction of the wale. Right-handed twills have the wale moving from the bottom left edge of a fabric up to the right. Left-handed twills are opposite; the wale moves from the bottom right edge of a fabric up to the left.

Twill fabrics have a face and a back. The face is the side of the fabric with the most pronounced wale. The face is usually more durable and is most often used as the fashion side of the fabric. Wherever there are warp floats on the technical face, there will be weft floats on the technical back. Sheer twill or printed twill fabrics are rare. Most twill woven fabrics are medium to heavy weight. The uneven texture of the twill weave hides soil better than any of the plain weaves. Twills are often used for upholstery because the interlacing pattern means that soils and stains are less noticeable. Fewer interlacings allow the yarns to move more freely so the drape tends to be better compared with plain woven fabrics. Also, with fewer interlacings, yarns can be packed closer together to produce dense, durable, high-count fabrics.

Wales in twill weaves can be more or less dominant depending on several factors. For more dominant wales, some combination of longer floats, combed or worsted yarns, plied or hard-twist yarns, and high counts are used. However, dominant wales are more subject to wear compared to more subtle wales. The twill line or wale may be steep, regular, or reclining. The angle of the wale depends on the balance; as the balance moves away from 1:1, the angle becomes steeper or more reclining. The greater the difference between the number of warp yarns and the number of weft yarns, the steeper is the twill line. Steep-twill fabrics are unbalanced in the warp direction (more warp yarns), but they are also stronger and more durable in that direction. Weft-faced and reclining twills are seldom used in furnishings because they are less durable than other twill-weave fabrics.

Even-sided or reversible twills have an equal amount of warp and weft yarn on each side of a fabric. Better-quality weft yarns are used in even-sided twill fabrics because both sets of yarn are exposed to wear. They are most often ^{2}—$_{2}$ twills. Upholstery twill flannel is often a ^{2}—$_{2}$ twill. Herringbone fabrics have the twill line reversed at regular intervals across the warp to produce a design that resembles the backbone of a fish, hence the name herringbone. Two different colour yarns may be used to accentuate the pattern. Herringbone patterns can be very subtle or very pronounced. Herringbone is used in furnishings for upholstery and draperies, and sometimes in table linens. Houndstooth is a ^{2}—$_{2}$ twill fabric with a unique small eight-point pattern. Two yarns in contrasting colours in the warp and weft are used in groups of four to create the distinctive pattern.

In warp-faced twills, warp yarns form the majority of the surface on the technical face of the fabric. Because warp yarns are more durable, these fabrics are stronger and more resistant to abrasion and pilling. Twill flannel and herringbone also can be warp-faced twills. Denim is a yarn-dyed cotton twill available in several upholstery and wall covering weights in a ^{2}—$_{1}$ or ^{3}—$_{1}$ interlacing pattern. Jean is a piece-dyed or printed medium-weight twill used for draperies and slipcovers. Covert is a ^{2}—$_{1}$ heavyweight twill upholstery fabric made of hard-twist worsted yarns that are fibre dyed or yarn dyed. Gabardine has a very prominent, raised, distinct wale that is closely set together. It is a warp-faced steep or regular twill. It always has many more warp than weft yarns. It can be made of carded or combed single or ply yarns. This durable upholstery grade fabric may be heather (fibre dyed), striped or plaid (yarn dyed), or solid (piece dyed) colour. Cavalry twill is another twill woven fabric with a pronounced steep, double twill line. The two diagonal wales are spaced very close together and slightly separated from the next pair of diagonal wales. Fancy twill interlacings that create fairly pronounced wales also are used because of their more interesting textures and patterns in upholstery, window treatments, and wall coverings. Broken twills are another type of fancy twill with a discontinuous wale and an interesting texture and pattern for use in furnishings. Goose-eye twill is one such example of a broken twill, with small diamond shapes developing due to a combination of twill interlacing, reverses in the interlacing pattern, and yarn colour. Institutional towelling is a twill weave fabric used in restaurants and foodservice facilities. It is usually white with one or two stripes of colour in the towel.

3.10 Satin woven fabrics

In most satin weave fabrics, each warp yarn floats over four weft yarns (most often designated as $^{4}—_{1}$) and interlaces with the fifth weft yarn, with a progression of interlacings by two to the right or the left (Fig. 3.7). In some more elaborate and expensive fabrics, each yarn floats across six or seven yarns and interlaces with the seventh or eighth yarn. Satin-woven fabrics are lustrous because of the long floats on the surface and are durable because of the very high warp yarn densities. In satin weaves, the yarns can be packed closely together, producing very-high-count fabrics. Although no two interlacings are adjacent to each other, satin weaves may look like twill weaves on the back. Careful examination may be needed to correctly identify the satin woven fabric structure.

Satin woven fabrics are available in two basic types; warp-faced where warp yarns cover the surface and weft-faced where weft yarns cover the surface. These fabrics are unbalanced with high counts that produce strong, durable fabrics with good body and firmness and resistance to soil. Because of the fewer interlacings, they are pliable and possess a firm, but fluid drape.

Satin woven fabrics are classified according to the yarn set that creates the surface: warp-faced or weft-faced. The term satin can refer to the weave or to a specific fabric. Satin fabrics are most often made with low twist fila-

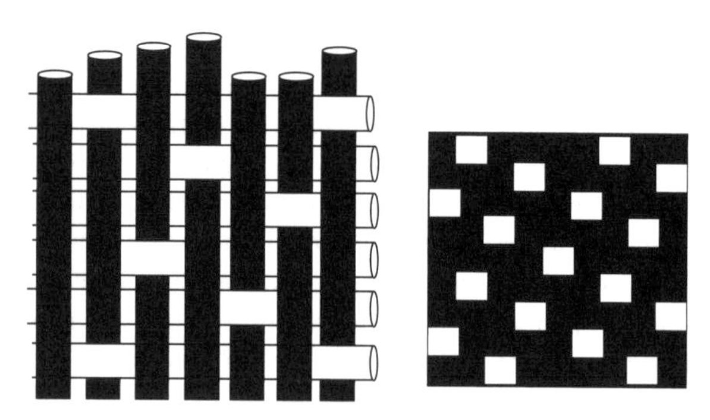

3.7 Satin weave interlacing (left) and checkerboard pattern (right).

ment yarns and are warp-faced. Because of the bright (not delustered) fibres, low twist, and long floats, satin is very lustrous. Its many weights are used in draperies, drapery linings, and upholstery. Crepe-back satin has crepe twist yarns in the weft that produce a soft and a luxurious drape. Antique satin has novelty yarns in the weft for more visually interesting upholstery and window treatment fabrics. Most sateen fabrics are weft faced with medium twist combed spun yarns. Finishes enhance sateen's lustre and durability. Weft sateen is a smooth, lustrous cotton fabric used for draperies and drapery linings. Warp sateen is made with warp floats to produce a stronger and heavier fabric. Warp sateen fabrics are often printed and used in bed sheets, pillow and bed ticking, draperies, and upholstery fabrics.

3.11 Fancy woven fabrics

In fancy weaves, the design, texture, and pattern are intrinsic to the fabric's structure and cannot be removed without dismantling the fabric. Some fancy fabrics are finished to more fully develop the design. Fancy fabrics are often more expensive than basic fabrics. The design in fancy wovens is created by changing the interlacing pattern between the design area and the background. There are many different ways of creating a fancy fabric.

Dobby weaves require fewer than 25 different warp arrangements to complete one repeat of the small, usually geometric, design. They are made on a dobby loom, a loom with a computer that controls the position of each warp yarn before insertion of a weft yarn. Bird's-eye has a small diamond-shaped weft-float design with a dot in the centre that resembles a bird's eye. Madras gingham has small, satin-float designs on a ribbed or plain ground. Waffle cloth has a three-dimensional honeycomb appearance. Waffle cloth is used for blankets and upholstery and for dish and bar cloths. Dobby fabrics are used for wall coverings, upholstery, and window treatment fabrics. Huck or huck-a-back is used for towelling. Its absorbency results from the short weft floats that form as a result of the dobby structure.

Jacquard weaves are large-figured designs requiring more than 25 different arrangements of the warp yarns to complete one repeat design. Fabrics made on an electronic jacquard loom with computer control of each warp yarn include damask, brocade, brocatelle, and tapestry. The differences among these fabrics are based on the weave structures that form the pattern or ground areas. Damask has satin floats on a satin background; the floats in the pattern are opposite those in the ground. This means that, when the pattern is warp-faced, the ground is weft-faced. Damask patterns have subtle, slight differences in light reflected from the two areas. Damask can be made from any fibre and in many different weights for upholstery, wall coverings, and window treatments. Damask is the flattest jacquard fabric and may be finished to maintain the flat look. Armure is a jacquard woven fabric with small motifs formed by warp yarns floating on the face while weft yarns float on the back of the fabric. It is not unusual for upholstery weight damask and armure to be constructed with warp and weft yarn sets that differ in lustre, degree of twist, and structure. Liseré is a slightly more complex jacquard fabric made with three yarn sets: two warp yarn sets and one weft yarn set. The weft yarn set and one warp yarn set are the same colour while the second warp yarn set is yarn dyed in several colours to form a coloured pattern of warp yarn floats in stripes on the surface. Frequently, the patterned stripes will be interspersed with satin stripes, creating a visually and texturally interesting surface. Brocade has satin or twill floats on a plain, ribbed, twill, or satin background. Brocade differs from damask in that the several colours of yarns that appear in the design are more varied in the length of their float. Brocatelle fabrics appear similar to brocade fabrics, but with a raised pattern. Brocatelle often has filament yarns with a warp-faced pattern and weft-faced ground. Coarse cotton stuffer weft yarns maintain the raised pattern area in the fabric when it is used for upholstery. Lampas is another complex fabric that combines satin and sateen weave interlacings to create motifs. Lampas usually combines two sets of a single colour warp yarn with one or more sets of multiple colour weft yarns to create a woven-in design of several colours. Jacquard tapestry is usually mass-produced for upholstery. It is a complicated construction of two or more sets of warp and two or more sets of weft interlaced so that the face warp never appears on the back and the back weft never appears on the face. Upholstery tapestry is durable if warp and weft yarns are comparable in size and structure. In lower-quality fabrics, fine warp yarns may be interlaced with coarse weft yarns, reducing the durability of the fabric. Some tapestry woven jacquard fabrics that resemble hand-made needlepoint fabrics are referred to as needlepoint weaves. In all jacquard fabrics made with multiple colours, the colours are present in lengthwise or crosswise bands and create coloured stripes on the technical back when they are not used on the face of the fabric.

Momie (mo'-mee) weaves have no wale or other distinct weave effect but an irregular interlacing pattern that looks as though the fabric has been sprinkled with small spots or seeds. Momie fabrics resemble crepe fabrics using yarns of very high or crepe twist. Some momie woven fabrics are variations of satin weave, with weft yarns forming the irregular floats. Other momie woven fabrics are even-sided while all others have a pronounced warp effect. Momie weave is also known as granite or crepe weave. Momie fabrics are used for upholstery, wall coverings, table linens, and window treatments. Sand crepe is a medium-weight to heavyweight fabric of spun or filament yarns with a large repeat pattern. Granite cloth is a momie weave variation of a satin weave. It is an even-sided fabric with

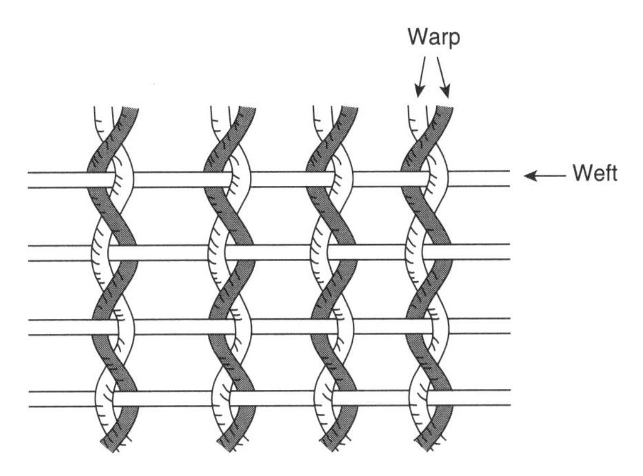

3.8 Leno weave.

no long floats and no twill effect. Moss crepe combines two-ply yarns in a momie weave. The yarns have only one ply that has a crepe-twist; the other ply may be filament or a medium twist spun yarn. Regular yarns may alternate with the plied yarns, or they may be used in one direction while the plied yarns are used in the other direction. Bark cloth is a heavyweight momie-weave fabric used for upholstery and draperies. The interlacing pattern is made using spun yarns that creates a rough, bark-like texture, thus giving the fabric its name. Bark cloth is available in prints or single colours. The rough texture adds interest and decreases the appearance of soiling.

In a leno weave, the warp yarns work in pairs; one yarn of each pair crosses over the other before the weft yarn is inserted (See Fig. 3.8). To create a leno weave, the loom must have a doup attachment to manipulate the warp yarns. When examining a leno fabric, it may appear that the yarns twist around each other, but that is not so. The warp yarns are crossed and one yarn of the pair is always above the other. The cross-over warp yarn is held in place by a weft yarn. Marquisette is a leno weave fabric used for sheer curtains and draperies. Casement draperies may combine a leno-weave with novelty yarns. Thermal blankets can be made of leno weave. All leno woven fabrics are characterized by open spaces between the yarns. The crossed-yarn arrangement produces a firmer, stronger, and more stable fabric compared to a plain-woven fabric of a similar low count.

In extra yarn weaves, supplementary warp or weft yarns of different colours or types are woven in to create a pattern. These extra yarns are wound onto a separate warp beam and threaded into separate heddles so that they are controlled individually and inserted into the design when required. When not used in the design, these yarns float across the back of the fabric and are most often removed during finishing. Many extra yarn woven fabrics have small-dot or geometric designs and are called dotted Swiss, clipped spot, clipped dot, or eyelash. They are used primarily for curtains, lightweight draperies and upholstery. Frequently, yarn colour is used to highlight the additional yarns and the pattern they create in the fabric.

The piqué (pee kay′) weave produces a fabric with wales or cords that are raised slightly by floats on the back. (While the term 'wale' is spelled the same for the vertical component in knits, the diagonal line in twill weaves, and the raised cords in pique, the three terms should not be confused with each other. They are not related other than that they refer to structural aspects of fabrics.) The wales vary in width and include wide-wale piqué (0.1 cm or 0.25 inch), pin-wale piqué (0.02 cm or 0.05 inch), other wale sizes and densities, and fancy piqués with raised designs. Stuffer yarns under the wales in better-quality piqué fabrics emphasize the roundness of the wale. The stuffer yarns are not interlaced with the surface yarns of the fabric and may be easily removed when analyzing a swatch of fabric. The cords or wales are usually parallel to the warp yarns. Bird's-eye piqué has a tiny design formed by stuffer yarns and the wavy arrangements of the cords. Bull's-eye piqué is a larger scale version of bird's-eye piqué. Both fabrics have crosswise cords and are used for upholstery. Bedford cord can also be made using the piqué weave. Bedford cord is a heavy fabric with raised wide warp cords used for bedspreads, upholstery, and window treatments. Stuffer yarns produce a more pronounced cord, all of one size or alternating larger and smaller cords.

A tapestry weave is a weft-faced, plain-woven fabric. True tapestry is hand-woven and frequently considered a work of art. Jacquard tapestry is a commercial imitation of true tapestry, but it is relatively easy to distinguish between the two. A true tapestry has discontinuous weft yarns arranged so that, as the colour or texture in the weave changes, a pattern is created. Discontinuous weft means that each weft yarn is found only in the area of the pattern requiring that colour or texture. Thus, unless a specific colour or texture extends across the full width of the fabric, the weft yarns change from one area to the next across the fabric. Each colour or type of weft yarn moves back and forth in a plain-weave interlacing pattern as long as the design calls for that colour or texture; then another colour or texture is used. In true tapestries, weft yarns are not always straight within the fabric and may interlace with the warp at some angle other than 90 degrees. If the change in colour or texture occurs along a vertical straight line, slits can develop in the structure. Different methods of structuring the fabric can enhance or eliminate the slit, depending on the effect desired by and the skill of the artist. True tapestries usually have larger weft yarns than warp yarns. Warp yarns are often covered completely by the weft, creating a weft-faced fabric. Some tapestries are designed to be displayed with the warp vertically while others are designed so that the warp is horizontal. If the tapestry is to be displayed on a wall, a backing may be needed for those tapestries that will have a horizontal warp in the finished piece. Historic tapestries found in museums and historic homes are often constructed in this fashion and require support to minimize damage from the weight of the piece. One-of-a-kind rugs, wall hangings, and fibre art pieces are made using this weave. Tapestry rugs include khilim rugs woven in Eastern European countries and Native American rugs. Some dhurrie rugs are a tapestry weave, but others are a plain or twill weave. Although pictorial wall-hanging tapestries are common, there are many categories of tapestries based on their pattern and end use.

Narrow fabrics encompass a diverse range of woven textiles that are up to 30 cm (12 inches) wide and made by a variety of techniques. Woven narrow fabrics include ribbons, window-blind tapes, piping used in upholstery and pillows, carpet-edge tapes, and trims. Webbings are narrow fabrics used in constructing upholstered furniture. Plain, twill, satin, jacquard, and pile weaves are used to create narrow fabrics.

3.12 Double cloth woven fabrics

The double cloth weave requires three or more sets of yarns compared to single cloth, which requires only one warp yarn set and one weft yarn set. The face and back of double-cloth fabrics usually are unalike because of the differences in yarn sets. Double cloths are heavier with more body compared to single cloths. There are three types of woven double-cloth fabrics: double cloth, double woven, and double-faced. Double cloth uses five sets of yarns to create two fabrics, woven one above the other on the same loom with the fifth yarn (warp) interlacing both layers. This technique is also used to produce velvet (see the section on pile weaves later in this chapter). True double cloth can be separated by pulling out the yarns holding the two layers together. It can be used in some upholstery fabrics. Although it can be confusing, double cloth is used to refer both to fabrics made with three or more yarn sets and to specific fabrics made with five yarn sets. Double cloth is more expensive and more pliable than single fabrics of the same weight because finer yarns can be used.

Double weave uses four sets of yarns that form two layers of fabric that periodically reverse position from top to bottom or face to back, consequently interlocking the two layers of fabric. Between the interlocking points, the two layers create pockets in the fabric. Double-weave fabrics are also known as pocket weave, pocket fabric, or pocket cloth. Most commonly seen in high-quality upholstery fabrics, double woven fabrics offer unique design possibilities, heavy weight, substantial body and drape, and good durability. Matelassé is a figured fabric made with either three or four sets of yarns. Two of the sets are the regular warp and weft yarns; the other sets are crepe or coarse cotton yarns. They are woven together so that the yarn sets crisscross. When the fabric is finished, the crepe or cotton yarns shrink, giving the fabric a puckered appearance. Heavy cotton yarns sometimes are used as stuffer yarns beneath the fabric face to emphasize the three-dimensional appearance of the fabric in upholstery weight matelassé. Cloque is made with four sets of yarns: two warp yarn sets and two weft yarn sets. The interlacing pattern is complex and simulates the appearance of quilting stitches on a solid coloured surface. Since there is no shrinkage of some of the yarns sets as in matelassé, cloque is a flatter fabric.

Double-faced fabrics use three sets of yarns: two sets of warp and one set of weft, or two sets of weft and one set of warp. Blankets, satin ribbons, and silence cloth are made this way. Expensive double-faced blankets of wool with one colour on each side are made with one set of warp yarns and two different coloured sets of weft yarns. In some blankets, patterns are created by interchanging the colours from one side to the other. Silence cloth is a white heavy cotton fabric that has been napped or brushed on both sides. It is used under fine tablecloths to absorb noise during dining.

3.13 Pile woven fabrics

Pile woven fabrics are three-dimensional fabrics made by interlacing an extra set of warp or weft yarns into the ground structure and creating short loops or cut ends on the surface. The pile is usually less than one cm (approximately 0.5 inch) in height. Woven-pile fabric is stiffer than knit or tufted pile fabrics. When a fabric is folded, rows of pile yarns may separate and allow the ground to appear or grin-through. When pile density is high, grin-through is less likely to be a problem. Pile fabrics are used for a variety of end uses: carpeting, upholstery, bedspreads, towels, window treatments, and decorative pillows. Attractive and interesting fabrics are created by combining cut and uncut pile, varying pile heights, mixing high- and low-twist yarns in the pile, intermixing areas of pile with flat surfaces, or flattening all or part of the pile. In these fabrics, the pile yarns receive most of the surface abrasion while the base weave or ground structure receives the stress. A strong and durable ground structure contributes significantly to a strong and durable pile fabric. A tight and dense ground weave increases the resistance to snagging for a loop or uncut pile fabric and the resistance of shedding or pile pull-out for a cut pile fabric. A dense pile is more likely to remain erect, resist crushing, and produce better cover. Careful cleaning will help keep the pile erect.

Pile fabrics are identified by the extra yarn set that is used to produce the pile – weft yarns or warp yarns (see Fig. 3.9). The pile in weft-pile

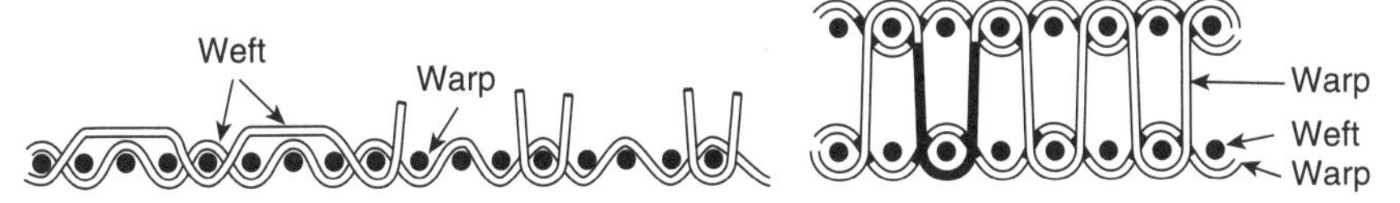

3.9 Pile weaves: weft pile (left) and warp pile, double cloth method (right).

fabrics is created by cutting long weft floats after weaving. Weft pile fabrics are always cut pile. Of the three yarn sets used to create a weft pile fabric, two sets are weft yarns and one set is warp yarn. The ground fabric is created using one set of weft yarns and the warp yarn set. In the weave structure, the extra weft yarns float across the ground yarns with many fewer interlacings. In corduroy, the floats are arranged in lengthwise rows and are less regular in velveteen. Cutting of the floats occurs in a machine, with metal guides that pick up the individual floats from the ground fabric and knives that cut the floats. In wide-wale corduroy, only one pass is needed. However, for pinwale or baby wale corduroy and velveteen, two or more passes through the machine are needed because the rows are so close together. After the floats are cut, the fabric is brushed crosswise and lengthwise to bloom open the cut ends of the pile yarns, raise the pile, and merge the pile into rows for corduroy or a uniform surface for velveteen.

Upholstery weight velveteen and corduroy usually use long-staple combed cotton yarns for the pile, and frequently for the yarns in the plain or twill weave ground. Twill ground weaves are preferred for upholstery because this structure produces a higher count and a denser pile. A dense pile produces a more durable upholstery fabric. Most corduroy has characteristic lengthwise wales, but no-wale or wale-less corduroy is also available. Velveteen has a staple fibre pile, more body, and is stiffer than velvet and has a more uniform pile surface than no-wale corduroy.

Warp-pile fabrics use two sets of warp yarns and one set of weft yarns where one set of warp yarns and the weft yarn set form the ground structure. The extra set of warp yarns forms the cut or uncut pile using one of several possible structures. The double-cloth method has been described earlier in this chapter. To make this structure, two fabrics are woven, one above the other, with an extra set of warp yarns interconnecting both fabrics. The two layers of fabric are cut apart while on the loom by a sharp blade or knife. With this pile method, the pile height is determined by the space between the two layers. The interlacing pattern of the pile yarns determines their resistance to shedding, density, and durability. Pile yarns can interlace with three weft yarns forming a W-shape as the pile yarn moves under one weft yarn, over the next weft yarn, and under the third weft yarn. Or, pile yarns can form a V-shape by interlacing with only one weft yarn. Because pile yarns with a W-shape interlace with more weft yarns, this shape of pile is more resistant to shedding, but it is less dense. Because pile yarns with a V-shape interlace with fewer weft yarns, the V-shaped pile is less resistant to shedding but it is denser. Determining the most appropriate pile shape will depend on the type of wear expected for the end use: surface abrasion, flex abrasion, or snagging. V-shaped pile is preferred for its better resistance to surface abrasion and flex abrasion while W-shaped pile is preferred for its better resistance to snagging. Pile shape can be determined by placing unravelled pile yarns on a surface of a contrasting colour and determining the shape of the fibres from the pile yarn.

Velvet is made of filament yarns with a short pile height (0.2 cm or less or 1/16 inch). Velvet is usually made with filament fibres and velveteen with staple fibres. However, wool or mohair velvet is available in upholstery weights. While many velvet fabrics are made in a single solid colour, variations are possible. Some velvet fabrics are made with a pile yarn of one colour and the ground yarns of another colour so that an iridescent effect is achieved. It is also possible to combine the pile structure of velvet with a jacquard weave to create a pattern with different colours or textures of pile yarns. In this structural variation, a fabric with a pile motif in one or more colours on a basic flat ground weave is created. Voided velvets are lush looking fabrics, but may not be as durable as velvets with a uniform pile due to variations in abrasion resistance between the pile and flat fabric areas.

Finishing is used to create other looks for velvet. Crushed velvet is made by mechanically twisting the wet cloth to flatten the pile in random directions. Panné velvet has the pile pressed flat in one direction to give it high lustre. If the pile is brushed in the opposite direction, the smooth, lustrous look is disturbed. Panné velvet is often made with rayon pile because of the bright lustrous look created by light reflecting from the flattened fibres. Sculptured velvets are created by pressing parts of the pile flat during an embossing step, shearing the fibres that remain erect, and steaming the flattened fibres so that they resume an erect stance. The steamed fibres are taller while the sheared fibres are shorter. Sculptured velvets are used for lighter use upholstery fabrics and formal draperies. Fur-like upholstery and decorative pillow fabrics may be finished by curling, shearing, sculpturing, dyeing, or printing any woven pile fabric to resemble furs of different species of animals.

In the over-wire method, a fabric is woven with wires placed across the width of the loom and positioned so that they are above the ground warp but under the pile warp. To produce cut-pile fabrics, each wire has a hook at one end with a sharp inner edge to cut the yarns looped over it as it is pulled through the loops. Uncut pile is produced using wires without hooks.

The wires are pulled through the loops before the cloth is removed from the loom. Velour, friezé, mohair plush, and most woven pile carpets and rugs are made in this way. Velvet can also be made by the over-wire method. Velour is a heavy warp-pile cotton fabric with a deep pile and more pronounced nap, used for upholstery and draperies. The pile in plush is usually at least 0.6 cm (1/4 inch) deep. Friezé can be a highly durable, elaborate pile fabric with only uncut pile loops or with both cut and uncut pile loops. It is an upholstery fabric that has fewer tufts per square inch than most other pile fabrics. The durability of friezé depends on the closeness of the weave and its fibre content. Complex patterns are possible by combining different-coloured yarns and loops and cut pile for velvet, friezé, mohair plush, and carpets and rugs. Grospoint is similar to friezé except that the loops are more pronounced and the surface is one level. Grospoint is highly durable and is frequently used for upholstery for office chairs and seat covers for buses and other heavy use vehicles.

The slack tension method is used to weave terrycloth. In this method, the pile is formed by inserting three picks or wefts before beating them into place with one motion of the reed. After the second pick in a set is inserted, the pile-warp beam relaxes some of the tension from the pile-warp yarns, while the yarns on the ground-warp beam remain at tension. After the third pick is inserted, all three picks are beaten firmly into place. As these picks move along the ground warp, they push the pile-warp yarns into small loops on one or both sides of the fabric. Most terrycloth is made with an all-over loop structure on both sides of the fabric. However, velour terrycloth has the loop structure on one side and a cut pile structure on the other side. Jacquard terrycloth has a woven-in pattern, created by varying the colour of the loop structure from one area of the fabric to another. Terrycloth is used for towelling and casual slipcovers. Friezé can also be made by the slack tension method. Another slack-tension fabric, shagbark, has widely spaced rows of occasional loops. Shagbark is sometimes used for window treatment fabrics.

Woven pile floor coverings include commercially woven velvet types, Axminster and Wilton techniques, jacquard types, and hand woven and hand knotted techniques. The woven pile carpets are complex structures with chain warp yarns, stuffer yarns to add body and weight, weft yarns, and pile yarns. The various yarns weave together to create the pile and backing layers in one simultaneous operation. The Axminster structure has paired weft yarns per pile row. Axminster rugs are more rigid in one direction and can be easily recognized because they roll or flex in one direction only. Axminster rugs are usually made to specific dimensions, often room-size, with an elaborate coloured pattern. Wilton rugs are a very heavy figured-pile fabric. In the Wilton structure, multiple colours of pile yarns are used to create the pattern. When these yarns are not visible on the surface to create the pattern, they are buried in the base structure, contributing weight to the rug, but also increasing its cost. Both of these types of rugs are expensive to weave and the patterns are often imitated by significantly less expensive tufting and printing techniques. Jacquard carpet is made with the jacquard attachment controlling the pile yarns. Velvet carpets are made using the over-wire method. Chenille carpets are sometimes considered a pile woven carpet, but the pile is due to chenille yarns with longer fibre tufts rather than incorporation of a separate pile yarn. Chenille carpets are relatively expensive carpets. Hand woven pile rugs were originally made in many parts of the world using a variety of hand knotting techniques. Each pile tuft in hand woven rugs is individually constructed using one of several knot types. As with most pile fabrics, pile density is a measure of the quality and durability of the finished carpet or rug. Examples include Oriental carpets, rya rugs, and flokati rugs.

3.14 Triaxial woven fabrics

A triaxial fabric is made with three sets of yarns. While it remains a woven fabric, the yarns interlace at 60° angles to each other rather than the 90° angle of the warp and weft yarns of the other woven fabrics. Triaxial fabrics are made with two sets of warp yarns and one set of weft yarns. In this fabric, the shedding arrangement creates a fabric with one set of warp yarns always passing over the weft yarns while the other set always passes under the weft yarns. This interlacing arrangement creates a lightweight and strong fabric with good elongation and resistance to tearing. Triaxial fabrics are sometimes used for support fabrics in upholstered furniture.

3.15 Nonwoven structures

Fabrics that are made directly from fibres or fibre-forming solutions are less expensive to make since there is no processing of fibres into a yarn. These operations include some very old and some very new processes. Fabrics made directly from fibres are referred to as nonwovens because they do not involve yarns. However, the term nonwoven can be confusing because knits are nonwovens as well. An alternate term is fiberweb. Nonwoven and fiberweb refer to a wide variety of fabric structures that all begin with a basic fiberweb. Nonwoven usage is increasing because of the increased cost of traditional textiles related to labour costs, costs of natural fibres, production and promotion of some manufactured fibres, easier cutting and sewing for unskilled labour, and new technologies that produce inexpensive made-to-order products. Nonwovens are used for disposable goods, such as kitchen and cleaning wipes and dust-cloths, as durable goods that are used to produce other products including dustcovers and support fabrics for furniture, as the base fabric for coated or laminated fabric for bonded or coated fabrics for upholstery, or as the fashion fabric for inexpensive draperies, office partitions or panels, and mattress pads.

Nonwoven or fiberweb structures include textile-sheet structures made from fibrous webs. These webs are bonded together by several means: mechanical entanglement of fibres, use of resins, thermal fusion, or formation of chemical complexes. Fibres are the fundamental units of the structure of the nonwoven fabric. The process is a simple one. The fibres are arranged into a web and bonded. Because of their similarity to paper, a key element is the distance between fibres. In nonwovens, the distance between adjacent fibres is several times greater than the diameter of the individual fibres making up the web. Because of this greater distance, nonwovens are more flexible than paper structures of similar construction. The properties of nonwoven fabrics can be manipulated fairly easily. Thus, it is important to consider the following factors when creating a nonwoven fabric: the properties of the fibres to be used in the web, the arrangement of the fibres within the web, and the properties of any binders or binding processes that may be used.

Production of nonwovens is rapid and inexpensive. Nonwoven fabrics of the same weight and fibre type as woven fabrics are approximately half the cost. The three basic steps in producing a nonwoven fabric are: selecting fibres appropriate for the end use, laying the fibres to make a web in the appropriate orientation and mass for the end use, and bonding the web together to make the nonwoven fabric. While any fibre can be used to make the web, the end use must be considered because fibre characteristics greatly influence fabric performance. Filaments and strong staple fibres produce strong and durable nonwovens; cotton and rayon contribute good absorbency; and thermoplastics are used to make spun-bonded webs. Five techniques are used in web formation: dry-laid, wet-laid, spun-bonded, spun-laced, and melt-blown. Fibre orientation refers to the direction of most fibres in the web. Fibre orientation contributes to many web characteristics. Thus, fibres can be arranged in a lengthwise orientation (referring to the machine direction or the direction in which the supporting conveyor belt moves), in a crosswise orientation (perpendicular to machine direction), or in a random orientation (fibres can point in any direction and are not parallel to each other). Oriented webs have the majority of the fibres parallel to each other. Random webs have fibres that are not parallel to each other. Lengthwise-oriented webs have a grain which must be considered when cutting pieces for production of products. Oriented webs are stiffer and stronger in the direction of orientation, but are softer and weaker in other directions. Random webs are more uniform in strength and stiffness in all directions.

Dry-laid fibre webs are the simplest type of nonwoven fabrics. They are made by carding or air-laying the fibres in a random or in an oriented arrangement. Carding is a fibre process that produces a parallel arrangement of short, staple fibres. Strength and pliability of the web can be manipulated by the number and orientation of layers of carded fibres used. One type of dry-laid web is a cross-laid web. This web is made by stacking one layer oriented in the lengthwise direction and the next layer oriented in the crosswise direction. Cross-laid webs do not have a grain so they can be cut more economically than woven or knitted fabrics. Another type of dry-laid web is the air-laid variety that has a random orientation. Air-laid webs are made by machines that disperse the fibres by air. While similar to the cross-laid web, this web has a more random fibre arrangement. End uses for dry-laid fibre webs include wipes, backing for quilted fabrics, building and wiring insulation, and base fabric for laminating and coating.

Wet-laid fibre webs are made from slurry that combines water with fibres of two lengths: short, paper-process length fibres and textile length fibres. The process of extracting the water leaves a randomly oriented and exceptionally uniform web of fibres. End uses for wet-laid nonwovens include laminating and coating bases for upholstery-weight and other supported films and other industrial end uses.

Spun-bonded or spun-laid webs are made immediately after fibres are extruded from spinnerets. The continuous hot filaments are deposited randomly on a fast-moving conveyor belt. Because they have not yet solidified and are in a semi-melted state, they fuse together at their cross points. They may be further bonded by heat and pressure. Because they are made from melt-spun synthetic fibres that are compact, strong, and lightweight, spun-bonded fibre webs have high tensile and tear strength and low bulk. Typical end uses for spun-bonded fibre webs include carpet backings, wall coverings, house-wrap vapour barriers, insulation, and roofing substrates.

Hydro entangled or spun-lace nonwovens are made by forcing jets of water through the fiberweb. The jets of water are under high pressure and hit the fibres from both sides of the fabric with so much force that they shatter the filaments into staple fibres, redistributing and entangling the fibres. The resulting fiberweb is a fabric that superficially resembles a woven fabric. The web has areas of greater fibre density and areas of lesser fibre density so that it looks like lace, hence the name of spun-lace nonwoven. These webs have greater elasticity and flexibility than spun-bonded fabrics. Because of the process, these fabrics are also known as water-needled fabrics. This technique makes products that are not possible with any other process. After the web is produced, the water is reclaimed, purified, and recycled. The degree of entanglement is controlled by the number and force of jets and the fibre type. Computer controls maintain a uniform quality in the fabric. Hydro entangled textiles are used in mattress pads, table linens, household wipes, wall coverings, and window-treatment components.

Melt-blown fibre webs are made by extruding synthetic heat-sensitive polymers through a single-extrusion orifice into a high-velocity, heated-air stream that breaks the fibre into short pieces. The fibres are collected as a web on a moving conveyor belt and are held together by a combination of fibre interlacing and thermal bonding. Because the fibres are not drawn, the fiberweb strength is lower than expected for a specific fibre. Olefin and polyester are used to produce inexpensive support fabrics for wall coverings and office panels or partitions.

Fibre webs become nonwoven fabrics through one of several processes: mechanical needling, use of polymers or adhesives, or application of heat. Mechanical needling or needle punching is done by passing a dry-laid fiberweb under a needle loom as many times as is necessary to produce the strength and texture desired for the nonwoven fabric. A needle loom consists of a bar with barbed needles protruding 5 to 8 cm (2 to 3 inches) from the base. As the needles stitch up and down through the web, the barbs pull a few fibres through the web, interlocking them mechanically with other fibres. Mechanical needling is a relatively inexpensive way to produce a thick and dense nonwoven fabric. Blankets, carpeting, and carpet backing are examples of needle-punched products. Fibre denier, fibre type, and product loft vary, depending on the end use. Indoor/outdoor needle-punched carpeting made of polyolefin is used for patios, porches, pool decks, and putting greens because it is impervious to moisture. Needled punched carpet backings are used with some tufted carpets.

Needled fabrics can be made of two or more layers of fiberweb, with each layer a different colour. By pulling coloured fibres from one layer to the top surface, geometric designs are possible. Needled fabrics can be made to resemble pile fabrics by pulling fibres above the surface. Needle-punched fabrics are finished by pressing, steaming, calendering, dyeing, and embossing. Embossing is used to create a three-dimensional texture in the nonwoven fabric. Other needling techniques include the use of a closed needle that penetrates the web, opens, grabs some fibres, and draws them back as a yarn-like structure that is then chain-stitched through the web. These fabrics are a type of stitch-through fabric. Needle-punched fabrics also are used for audio speaker-cover fabrics and synthetic leathers.

Chemical adhesives, polyvinyl acetate, vinyl/acrylic, and acrylic polymers, are used to bond the fibres together in dry-laid or wet-laid webs. Each adhesive type has characteristics that make it appropriate for certain applications. The appropriate adhesive is applied as a liquid, powder, or foam, heat-activated, and pressed to adhere fibres together in the web to form the nonwoven fabric.

Several techniques requiring heat and pressure are used to bond thermoplastic fibre webs. In area-bond calendering, the fibres are heated and pressed to form a stiff, permeable film-like structure that is non-elastic and strong. In point-bond calendering, the fibre web is heated and passed through one pair of engraved and smooth calender rollers. The engraved roller presses the web onto the smooth roller to force the fibres to adhere to each other in the pattern of the engraved roller. Characteristics develop based on the size and frequency of the bonding pattern. Fabrics are moderately bulky, elastic, and soft, and are used for wipes and filtration applications.

3.16 Fiberfill

Even though batting, wadding, and fiberfill are not fabrics, they are important components in furnishing textiles for quilts, comforters, furniture padding, pillows, mattresses, and mattress pads. Each is made from fibre, but differences exist among the three. Batting and fiberfill are made from new fibre, wadding is made from waste fibre. Fiberfill is further differentiated from batting because it is a manufactured fibre staple made especially for use as a filler. Batting is usually made from lower quality natural fibres that are not well suited for processing into yarns. Most often, batting is cotton. Batting fibre is lower quality than yarn quality fibre because of its short length, irregular colour, or because it is too immature and will not process well into yarns. Batting often includes recycled fibre from bale opening or yarn processing facilities.

To make fiberfill from staple fibres, the fibres are carded and laid down to the desired thickness and may be covered with a sheet of nonwoven fabric. Fibre density describes the weight or mass per unit volume. Fibre density is an important factor related to end use requirements for resiliency and weight. Resiliency is important because fabrics that maintain their loft incorporate more air space. When fibres stay crushed, the fabric becomes thinner and more compact, losing bulk which decreases its ability to insulate and pad or shape furniture and other soft furnishings such as window cornices. Resistance to shifting means that a fiberfill will maintain a uniform thickness. For instance, down comforters need to be shaken often, because the down shifts to the outer edges. Thermoplastic-fibre batting can be needle punched to minimize shifting. In this process, hot needles touch and melt portions of some fibres and fuse the fiberfill into a more stable batting. This finished product is a bonded web. The ability of a batting or fiberfill to insulate is based on two factors: the amount of still air trapped in a volume of fabric and the number of fibres present. In some furnishings where a low profile is desired, micro fibre fiberfills are popular. Because of issues related to recycling fibres, several fiberfills are made from recycled polyester solutions derived from beverage bottles and other plastic containers.

Polyester fiberfills are available in several modifications. The high-loft type maintains a bulky profile during use. The hollow fibre type includes voids that increase loft and insulation with minimal additional weight. The microfibre varieties shape with less bulk and better drape. *Primaloft* by Albany International Research Company is a microdenier polyester fiberfill that mimics down in weight and loft, and is used in bedding and pillows.

3.17 Down and feathers

Down and feathers are also used as fiberfill. These fills are defined by various governmental agencies to control labelling and commerce in less expensive imitations sold as the real item. Down refers to the fine, bulky under-plumage of waterfowl (ducks or geese). Items with labels that read '100 percent down' must meet specified requirements for the condition of the down. Down-filled items must be 80 percent down or down fibre and may include up to 20 percent other feathers. Down is also rated by its loft capacity or fill power – the volume that one cubic ounce of down will fill. For example, 650 down is warmer and more expensive than 550 down. The loft capacity range is generally 300 to 800. Down is lightweight and warm. However, it is likely to shift with use, and when wet, it mats and loses its warmth. It is difficult to maintain down's original loft, and it is difficult to clean. People with allergies to feathers may have problems with down products. Down is used in bedding and padding for pillows and soft furniture. Feathers are much less desirable as a fill agent compared to down. They are, however, substantially less expensive.

3.18 Felt

True felt is a mat or web of wool, or mostly wool, fibres held together by the interlocking of the wool scales. No yarns are present in felt. It is made directly from fibres and is one of the earliest types of fabric. Felt can be made by washing wool fleece, spreading out the wet fleece, and beating or handling it until it mats and shrinks together in a flat or three-dimensional shape. Commercial felt is made in factories by layering wool fibres or blends of wool and other fibres until the desired thickness is attained, using heat, detergent, and vibration to mat the fibres together, and finishing the fabric. Felt finishing resembles the processes used for woven fabrics. Some fabrics labelled 'felt' are not true felt because they do not include wool. Instead, these imitation felts are wet- or dry-laid nonwoven fibre webs. Felts do not have grain and do not ravel. They are stiff, less pliable, and weaker compared with other wool fabrics because they lack yarn structure, which contributes to fabric pliability and strength. The quality of felt depends on the quality of the fibres used. Felt is used in hand-made wall hangings, other types of ornamentation, and party and holiday decorations.

3.19 Leather, suede and fur

Leather, suede and fur are produced from skins and hides that are processed to maintain flexibility after being removed from the animal. The fibres in leather are very dense on the skin side and less dense on the flesh side (see Fig. 3.10). Thick hides are often split or shaved into layers to make them more pliable and economical. Mammal, reptile, bird, and fish hides and skins are processed into leather. In order to use the skin, the animal must be killed. Most leather is a by-product of the meat industry. While animal-rights activists find that killing animals for their skin to be cruel and unnecessary, many others find leather, suede, and fur attractive and desirable regardless of their source.

Product design is limited by the size of the animal. 'Skins' are usually removed from smaller or younger animals such as lambs, calves, and goats and are usually less than one square meter in size. Skins are usually thinner

3.10 Cross section of leather.

than hides. 'Hides' are removed from larger and older animals including cattle, deer, and horses and are usually less than three square meters. Because most furnishing applications for leather require heavier and larger pieces, hides are usually used. Because of the size limitations imposed by the type of animal from which the hide is removed, leather items may have more seams compared to similar items made of other materials. Leather is used for upholstery, wall coverings, wall hangings, and accessories such as lamp shades and coasters.

Leather quality is determined by the environment, genetics, and processing. Environmental issues include exposure to factors that can permanently scar a skin or hide. Genetics refers to the species as well as the breeding of a specific animal. Processing refers to how the hide or skin has been handled from removal to making up into a finished product. The hides and skins from different animals vary greatly in uniformity in size, thickness, and grain. Grain refers to the markings that occur during growth and ageing of the skin. Grain varies from one species to another, from one animal to another, and within a single hide. Ageing and other conditions influence the surface of hides. Permanent scars develop as animals fight with each other, cut themselves on fences or thorns or nails, and are bitten by insects. Brands, tattoos, and skin diseases also permanently mark hides. Vein marks, wrinkles, or pronounced grain marks are undesirable irregularities. Because of the frequencies of these flaws, fewer than 5 percent of hides are converted into the best quality leather: smooth top-grain leather in aniline finish. Approximately 20 percent of leather is made into smooth leather with a pigment finish. The remaining 80 percent is corrected in finishing so the flaws are not obvious. These processes help produce a leather of a consistent or uniform look.

Leather is a durable product. It should not have a noticeable odour. It varies greatly in quality – not only from skin to skin but within a single skin. Leather from the backs and sides of the animal is denser and of better quality than leather from the belly and legs, which can be thin or coarse. Leather picks up oils and grease readily. Removing soil without creating other problems such as stripping colour, reducing pliability, making flaws and blemishes visible or removing or altering screen prints or embossed grain usually requires the hiring of a specialist in cleaning leather.

Dried skins and hides as they are removed from a carcass are stiff, boardy, non-pliable, odorous, and subject to decay. They go through many steps in the process of becoming leather. They are defleshed and washed to clean the hide and replace moisture that was lost, especially necessary with hides that have been salted and stored for months. (Salting is used to keep the hide from rotting before it can be processed.) Cleaning removes the hair and the epidermis. Pickling is a treatment with acid, salt, and water to swell the tissue and prepare it for tanning. In tanning, the skins and hides are treated with chemicals to make them rot resistant, pliable, and water resistant. Alum tanning makes white leather. Oil tanning is used to make chamois. Vegetable tanning uses a bark extract and is the most expensive process. Chrome tanning uses a solution of bichromate of soda, sulphuric acid, and glucose. Because of environmental concerns with the highly toxic waste from chrome tanning, vegetable tanning is gaining in popularity. A new tanning process using fluorochemicals to cross-link the collagen in the skin produces a leather that is waterproof, shrink- and stain-resistant, breathable, washable, and soft.

Stretching and drying ensure that the leather will have elasticity so that it gives with use and recovers from stretching. This means that seats of leather upholstered furniture should not acquire an inflated look after use. Sammying squeezes out excess moisture before additional processing. Hides at this intermediate stage are called wet-blue. Because mature animal skin can be very thick, many hides are split to less than 1–2 mm (0.1 inches) thick. Shaving is done to provide a uniform thickness. Fat-liquoring, or stuffing, adds oil to the leather so that it remains flexible after drying and maintains a soft hand. Full aniline leather is leather with a transparent dye that contains no pigments. Pigment finished leathers are usually stiffer than aniline dyed leathers because of the application to the surface of the leather of three-to-four coats containing pigments, binding agents, and other materials. Pigment finished leathers are usually abrasion and stain resistant. After drying, leathers are stiff and boardy. Staking is a mechanical process that flexes the hide and makes it soft and pliable again. Buffing is a mechanical sanding process that smoothes any blemishes on the hide. Snuffing is a slight abrasive treatment of a hide to camouflage minor flaws. Hides that have been buffed or snuffed are known as corrected grain. Embossing or plating produces simulated grain patterns on the surface of the leather. Glazing produces leather with a bright, shiny surface that usually has enhanced resistance to moisture because of the waxy, resin, or lacquer-based compounds used in the glaze.

The first or outermost layer of leather is the top grain because it has the characteristic animal grain. This leather takes the best finish and wears well. It is also the highest quality and most expensive. Full-grain leather has not been cosmetically altered in any way compared to top-grain leather that may have had some minor cosmetic adjustments made. Split leather has a coarser and more porous structure that may become rougher with use. Most split leathers have an embossed or suede finish. Although splits may not be identified on product labels, top grain is usually mentioned because of its higher price and quality. As more splits are removed from a hide, the quality of each split decreases layer by layer. The lowest-quality split leather is next to the animal's flesh. Full-grain leather is not split.

Leather may be sold by thickness or weight. When thickness is measured in millimetres (mm), divide by four (4) to determine the weight in ounces. However, weight is not necessarily an indicator of quality since many other factors influence that.

Suede is a leather product used for upholstery, decorator pillows, wall hangings, and wall coverings. The soft, dull surface is made by napping or running the skin under a coarse sander on the flesh side or on one side of a split to pull up the fibres. Nubuck leather is napped on the grain side to create a soft hand. Moisture can damage the texture and appearance of any suede. Cleaning of suede should be done by a specialist.

Fur is any animal skin or part of an animal skin with the original hair, fleece, or fur fibres attached. Some fur is used for throws, rugs, decorator pillows, and wall hangings. Furs are natural products and vary in quality. Good-quality fur has a very dense pile with a pliable, very soft hand. When a fur has guard hairs, they should be long and lustrous. Fur quality depends on the age and health of the animal and the season of the year in which it was killed. Furs are usually harvested in late fall because the coat is most dense with minimal sun and abrasion damage.

For centuries, fur trapping was important in all parts of the world. Fur farming, begun in 1880, has resulted in better quality pelts because of scientific breeding, careful feeding and handling of the animals, and slaughtering them when the fur is in prime condition. Silver fox, chinchilla, mink, Persian lamb, and nutria are commonly raised on ranches. The cost of fur depends on fashion, the supply and demand for fashionable furs, and the work involved in producing the item. Chinchilla, mink, sable, fox, and ermine are usually the most expensive furs. Skins are gathered from sources worldwide and sold at public auction.

Furs go through many processes before they are sold as products. The dressing of fur is comparable to the tanning of leather, except that more care must be taken so that the fibres are not damaged. After tanning, pelts are combed, brushed, and beaten. The final process is drumming in sawdust to clean and polish the hair and to absorb oil from the leather and fur. The sawdust is removed by brushing or vacuuming. Many furs are dyed to improve their natural colour, to make them look like more expensive furs, or to give them an unnatural colour such as red. Other dye processes include tip-dyeing (brushing dye on fibre tips) and dip-dyeing (dipping the entire skin in dye). Some furs are bleached or bleached and then dyed.

Furs require special care. Do not use fur in humid places or store fur accessories (such as pillows) in plastic bags. Protect furs from exposure to insects and direct sunlight that will dry out and bleach the fibres. When not in use, store fur accessories in temperature- and humidity-controlled vaults (usually a service provided by fur retailers). To clean fur, use a specialist who utilizes the furrier method in which the fur is tumbled in an oil-saturated, coarsely grained powder that absorbs soil and lubricates the fibres and skin with oil while exposing the fur to minimal abrasion and matting. After tumbling, items are brushed and vacuumed to remove the soiled powder.

3.20 Future trends

Future trends in textile construction include an increasing interest in green or sustainable design, which is a focus on how design goals are achieved and their effects on people and the environment. Textile production has been sensitive to environmental design for years. Examples include changes in the types of machines used to produce fabrics, an ever-increasing emphasis on fabric quality, and changes in how yarns and other materials are prepared before being made into fabrics. Newer looms reduce environmental impact. Water-jet looms require clean recycled water to carry the weft yarn across the fabric. However, fabrics produced with water jet looms must be dried before storage to reduce problems with mildew. Drying fabric uses large amounts of energy. Even though today's looms are much faster than traditional shuttle looms, energy use varies significantly with the type of machine. Projectile looms are preferred over rapier and air jet looms because of their lower energy use.

Recycled synthetic textiles are becoming more readily available, including those from recycled carpets and beverage bottles. Recycled natural fibre textiles, while not as widely used as recycled synthetic fibres, are also available.

For woven fabrics and most knit fabrics, yarns are treated with chemicals to minimize problems with abrasion during fabrication. These chemicals are usually removed after the fabric has been produced and are often reclaimed. Lint resulting from yarn abrasion during fabrication can create fabric quality problems. Vacuum heads attached to flexible tubing remove lint and minimize health, quality, and equipment problems. Producing better-quality fabric improves efficiency and decreases environmental impact. Fewer fabric flaws mean less waste and fewer seconds from cut-and-sew production facilities.

Reducing production costs has been a long-term strategy in the textile industry. Most textile production occurs in South and Central America, Asia, and some parts of Africa because of low labour costs and fewer environmental restrictions. Because so many companies have moved textile production facilities to other parts of the world, more emphasis on specifications and standards are necessary to ensure that the goods received are of the required materials and quality and meet performance expectations. Unfortunately, this may create challenges in terms of sustainable design expectations that materials are manufactured locally.

Another trend is towards shorter production runs of basic and designer fabrics. This trend allows for more fabric options in the market. Computer use will continue to provide relatively inexpensive options and increased colour assortments for printed and dyed fabrics, and structural design fabrics.

3.21 Sources of further information and advice

There are numerous trade publications, magazines, and journals that provide current information on textile structures. *Home Furnishings News*, *Interior Design Magazine*, and *Interiors and Sources* are three examples.

There are also several websites provide useful information. The American Society of Interior Designers (ASID) (http://www.asid.org/) is an organization for interior designers. The website has numerous pages with information about interiors, interior design, starting and marketing an interior design business, and related topics. They publish a magazine, ASID ICON, six times a year that includes design research and business trends. The website includes a knowledge centre that provides information about publications, products, codes and standards, research, and topical issues. Current information about textiles is included in the magazine and on the website.

The International Interior Design Association (IIDA) (http://www.iida.org/) is a professional networking and educational association, and includes thousands of members from around the world. IIDA resulted from the merger of the Institute of Business Designers (IBD), the International Society of Interior Designers (ISID), and the Council of Federal Interior Designers (CFID) in 1994. As with ASID, information about textiles is included on the website.

3.22 Bibliography

The following publications deal in more detail with some of the topics discussed in this chapter:

EARNSHAW P (1982), *A dictionary of lace*, Aylesbury, UK, Shires Publications.

EMERY I (1980), *The primary structures of fabrics*, Washington, D.C., The Textile Museum.

GRAYSON M (1984), *Encyclopedia of textiles, fibres, and nonwoven fabrics*, New York, Wiley.

HUMPHRIES M (1996), *Fabric glossary*, Upper Saddle River, NJ, Prentice Hall.

JACKMAN D R, DIXON M, and CONDRA J (2003), *The Guide to Textiles for Interiors*, Winnepeg, Canada, Portage and Main Press.

JERDE J (1992), *Encyclopedia of textiles*, New York, Facts on File, Inc.

KADOLPH S J (2008), *Textiles*, 10th ed., Upper Saddle River, NJ, Pearson Education.

KADOLPH S J (2008), *Quality assurance*, 2nd ed., New York, Fairchild Publications.
NELSON K J (2007), *Interior textiles: Fabrics, application, and historic style*, New York, Wiley.
PARKER J (1992), *All about silk*, Seattle, WA, Rain City Publishing.
PARKER J (1996), *All about wool*, Seattle, WA, Rain City Publishing.
PARKER J (1998), *All about cotton*, Seattle, WA, Rain City Publishing.
SCHWARTZ P, RHODES T, and MOHAMED M (1982), *Fabric forming systems*, Park Ridge, NJ, Noyes Publications.
SPENCER D J (1983), *Knitting technology*, New York, Pergamon Press.
TORTORA P G and MERKEL R S (1996), *Fairchild's dictionary of textiles*, 7th ed., New York, Fairchild Publications.
YATES M (1995), *Textiles: A handbook for designers*, New York: Prentice Hall.
YEAGER J I and TETER-JUSTICE L K (2000), *Textiles for residential and commercial interiors*, 2nd ed., New York, Fairchild Publications.

4

Surface design of fabrics for interior textiles

T. DAS, Dicitex Decor Pvt. Ltd., India

Abstract: This chapter aims to provide a comprehensive understanding of the elements and techniques of fabric surface design in the context of interior textiles. The discussion starts with an introduction of the basic surface design elements along with a historical perspective of the subject. This is followed by a description of the techniques of dyeing, printing, creating textures and embellishing textiles. The roles of designers and trend forecasters in the industry are also explained before the chapter concludes with a discussion on the future of interior textile design.

Key words: fabric surface design, interior textiles, dyeing, printing, trend forecasts.

4.1 Introduction

Surface design of textiles constitutes the appearance of the fabric surface in terms of its colour, texture, and, if applicable, pattern. The surface of the fabric is an integrated part of the whole, especially in the case of interior textiles where the performance, durability, and other functional requirements of the fabric is of extreme importance. Consequently, the surface design should be seen, not just as an aesthetic packaging of the product, but also as a summary of the textile's quality; in other words, a convergence of the effect of all the finishes and treatments that went into making the textile with its visual qualities.

The aim of this chapter is to provide a comprehensive understanding of the elements and techniques of fabric surface design while keeping in perspective the history, as well as the future, of this subject in the context of interior textiles. In that direction, Section 4.2 introduces the basic surface design elements and their significance in interior textiles. Section 4.3 provides a historical perspective of interior textiles and their evolution in the West, before we move on to a description of the basic surface design techniques of dyeing, printing, texturing and embellishing fabrics, in Sections 4.4–4.7. Section 4.8 briefly touches upon some innovations and advancement of surface design techniques. In Section 4.9, the role of design forecasts and trends in the design industry are discussed. Finally, the chapter ends with some speculations, in Section 4.10, about the future of interior textile design.

4.2 Fabric surface design – Its significance in interior textiles

The significance of the design elements of colour, texture and pattern, of an interior textile, depends on the context in which it is used in the interior environment. The following sub-sections briefly discuss some specific requirements of interior textiles with respect to these elements. The section ends with a few definitions of surface design terms.

4.2.1 Colour

The colour of any interior textile plays one of the most dominating roles in determining its marketability. Seasonal trends will always significantly influence the consumer and the designer's choice of colours. But there is also a range of other factors that determine the applicability of colours within the interior. For window treatments, for example, the choice of colour would depend on the intensity, angle or glare of daylight received, and also the kind of artificial light used within the room during night time. For a sheer fabric used to control glare from windows that are flooded by daylight, a dark colour will be preferable. On the other hand, for places that receive less sunlight, use of bright colours will help to brighten up and also tint the light filtering in. In the case of upholstery, the choice of colour would depend not only on the character or structure of the piece of furniture in question, but also on the other colours already existing in the room.

4.2.2 Texture

The texture of a fabric evokes visual as well as tactile response from the user. For an upholstery fabric or for soft floor coverings, this can be used to advantage by making the surface appear soft, seductive or luxurious. At the other extreme, the texture should not be coarse to the extent that it would feel unpleasant to the touch, or lead to mechanical problems such as snagging, fuzzing or pilling. For sheer curtains, interesting textures formed by the use of novelty yarns in a background of open structure, can create a beautiful silhouette with the light filtering in.

4.2.3 Pattern

The *pattern* of a textile is the visual arrangement of elements and motifs on its surface, a character that distinguishes it from any other surface. The pattern might be a result of the fabric construction, such as the knit or the weave, or it may be printed or painted on the surface in colour and/or texture. The pattern might be a random arrangement of motifs or abstract elements, or a formally arranged repeating unit. The *pattern repeat* is one complete unit of the surface design. A flawless pattern results in an unbroken flow of the design when tiled in all directions, giving no impression of unintentional lines in any direction when viewed in totality. The *motif* is a single element in a pattern, for example a flower, or a dot or a vine.

Within a given space, it is desirable for the scale of the pattern to be in harmony with other elements in the room, and also the size of the space in consideration. Large patterns with movement usually look good when used in window treatment, but (with exceptions) might be a wrong choice for upholstery. In general, patterns used on upholstery fabrics need to be balanced, or structured, so as not to conflict with the character and form of the furniture itself.

4.3 Textile design – A historical perspective

4.3.1 Early times

The association of aesthetics with textiles has led to the pursuit for colouring and decorating fabric since time immemorial. The earliest known attempts at *dyeing*, presumably from the Neolithic Age, involved colours extracted from minerals and plants that could produce only temporary stains on fabrics (Wells, 1997). In ancient Egypt and India, indigo was extracted by fermenting the crushed leaves of the indigo plant to produce indigo blue. Shades of brown, yellowish brown and cream were obtained from tannin-containing substances, such as the barks, stems and seeds of certain plants. The rare purple dye, extracted by the Phoenicians from the shellfish of the Mediterranean Sea at Tyre, was so expensive that it could be used only by the nobility. Other popular sources of colour included the root of madder, the leaves of henna, and the petals of safflower that yielded reds, oranges and yellows. The discovery of mordants; later during the medieval ages in Europe, increased the permanency and versatility of colours obtained from leaves and flowers. (Storey, 1978)

The art of applying patterns on fabric by *printing* dates back to at least 3000 B.C., as evidenced by some early wooden blocks that have been found. Block printed and resist dyed (tie-dyed) cotton fabrics from India became an extensively traded commodity throughout Asia and the Mediterranean region after Alexander's invasion of India in 327 B.C. Indian calico and chintz were again popularized in Europe by the Arab traders around the second century A.D., and by the British East India Company during the seventeenth century. In parallel with India, block printing existed in China for a long time (Wells, 1997). Intricately printed Chinese cotton fabric and Chinese motifs became very popular in Europe, especially in France, during the 'Rococo period' in the first part of the eighteenth century. In 1752, copper plate printing, invented by Francis Nixon in Ireland, enabled a finer resolution and larger repeat of printing. By 1797, the benefit of this invention was fully reaped by Christophe-Philippe Oberkampf in his factory in Jouy-en-Josas in France. The printed cotton toiles of excellent quality produced at this factory, came to be known as 'Toile de Jouy' and are still popular for their distinctive style (Lebeau, 1994). Resist printing and dyeing are other patterning techniques that were developed in different parts of Asia and also Africa, e.g. ikat, batik, shibori or plangi.

4.3.2 Evolution of soft furnishings in Europe (A.D. 1600–1900)

Until the *Medieval* era, there was generally no demarcation between upholstery and apparel fabrics in the West (Lebeau, 1994). During the English Elizabethan (A.D. 1558–1603) and subsequent Jacobean era (1603–1625), a period of relative stability in England, people could start to afford some domestic comfort. Interiors that were previously bare and hard began to acquire some furniture and fabric. Furnishings were in wool, silk, linen or leather. Crewel embroideries on wool and linen became very popular. Furniture upholstery was not yet known.

During the *Renaissance* era in Europe (fifteenth to the mid-seventeenth century), interiors assumed a more formal look, incorporating some ornate classical elements of decoration, under the influence of architects such as Andrea Paladin (Nielson, 2007). Florence, Lucca, Venice and Genoa became great centres of silk weaving, producing beautiful satins, damasks, brocades, brocatelles, lampas, and velvets, richly patterned, in vivid colours. Genoese velvets, whether plain, multicoloured, goffered or patterned, acquired a class of their own. Beds continued to be more elaborate, with ornately carved headboards, canopies, posts, cornices, pelmets and valances. Bed draperies went all around or were tied back at the bedposts with tassels and trimmings.

The luxuriousness of the Renaissance period continued into the *Baroque* era (A.D. 1580–1750), with an even greater ostentatiousness. Textiles had bigger motifs in brighter hues, and had gold and silver threads woven into them. Floral bouquets, trellises, acanthus leaves and symbols of the Sun King, were some of the popular motifs. Bed and window draperies had elaborate embroidery, passementerie and lace work. The Savonnerie and moquette rugs were signature Baroque furnishing items. Lyons, in France, became a major silk-producing centre under the influence of Louis XIV (Lebeau, 1994). The Chateau de Versailles beautifully preserves some of the grandeur of the Baroque era.

Interior aesthetics changed into a more feminine, dainty, and elegant style, under the influence of Madame de Pompadour and Madame du Barry, during the reign of Louis XV in France, called the French *Rococo* period (A.D. 1730–1760). Motifs had curved lines and floral patterns, projecting a romantic garden theme. Colours used were lighter compared with the earlier periods. 'Chinoiserie', or the influence of Chinese aesthetics, became very popular. Interiors were smaller and cosier, with lighter furniture pieces outlined in wood carved with such Rococo motifs as shells, and upholstered at the backs, seats and arms. Beds had upholstered headboards and footboards and some, called the 'lit clos', were built sideways into a closeted space with side draperies. Windows were draped in heavy, tie-back draperies and sometimes had another layer of sheer curtains beneath. Aubusson and Savonnerie rugs were favourites as floor covering (Nielson, 2007).

The interior aesthetics that followed in Europe from the mid-eighteenth to the mid-nineteenth century, called the *Neoclassic* and *Empire* periods, were heavily influenced by the classic Greek look, unearthed at Pompeii during the excavations of 1754. Under the influence of the Scottish architect Robert Adam, colours were lifted to creamier, pastel values. The advent of copperplate printing and the manufacturing of machine-made woven pile carpets were significant contributions of this time. The cord and pulley system for drawing curtains was also a development of this time. The Neoclassic style continued into the *Empire* period (1804–1820) during the reign of Napoleon Bonaparte, stamped heavily by his personal choice and preferences. These preferences included loosely draping fabrics on walls to create the effect of military tents, inclusion of dark, bold colours in the palette, creating rigid looking furniture, and the use of imperial emblems (Nielson, 2007). The advent of the Jacquard and the development of roller printing were some revolutionizing changes in textiles from the Empire period.

The nineteenth century ended with the *Victorian* era (1837–1901). As industrialization and mass production continued during this period, there was a gradual phasing out of hand-made textile products and an increasing abundance of machine-made woven and printed products in 'revival' styles, borrowed from previous historic periods and styles, such as Gothic, Islamic, Neoclassic, Rococo and so on. With the invention of chemical dyes by William Perkins, a wider range and greater choice of colours for dyeing, became available (Storey, 1978).

4.3.3 Modern history of interior textiles in Europe and the USA

Both the *Art Nouveau* (1890–1910) and the *Arts and Crafts* (1905–1929) movements in Europe and America, started as a revolution against industrialization and the loss of creative satisfaction inherent in mass production. Everything handmade was encouraged. Scottish architect Charles Rennie Mackintosh, French designer Hector Guimard, Spanish architect Antonio Gaudi and American glassmaker Louis Comfort Tiffany are only a few of the outstanding designers of this period. Art Nouveau designs and motifs were inspired by nature and foliage, and characterized by graceful, stylized, movement and rhythm. Furniture was constructed with simple, exposed, structural lines. One of the most influential designers of the Arts and Crafts movement was William Morris. His designs for wallpaper and textiles were characterized by complex layers of curves with a lot of movement. The complexity of these patterns was in direct contrast with the simplicity in form and structure of the furniture of those times – especially those designed by Gustav Stickler (Nielson, 2007).

The developments in art and design during the *Art Deco* period (1918–1945) were in two directions. One was influenced by the Beaux Arts Paris designers and was characterized by the application of sleek, stylized decorative motifs inspired by machines, speed, geometric designs as found in Egyptian pyramids and Aztec architecture, Cubism and Futurism. The other direction was influenced by the minimalist, uncluttered, and functional approach of the International Modern designers. Neutral, grey and earthy colours, sometimes stylized with bright accents, were in vogue.

The *International Modern* style of design that started since 1932 and continues to this day was inspired by the 'De Stijl' and the Bauhaus movement, the Scandinavian Modern style, and philosophies of architects such as Le Corbusier. The underlying objective of this school was to create affordable designs based on the ideals of 'functionality and purpose over form'. The minimalist Japanese interior and the structurally exposed modern Scandinavian style of construction were considered as perfect fits to the International Modern school of thinking. Furniture was simple, as was the upholstery. Another very important and profound influence on modern interiors and textiles is the *organic design* philosophy of *Frank Lloyd Wright* (1867–1959). Wright, through his organic style, aimed at attaining a perfect harmony between nature, landscape and his creations (Nielson, 2007).

This section has discussed the changes and highlights of interior design and textiles through the various periods from the medieval era to the present, as witnessed in Western Europe and in some cases the USA. The evolution of interior design in each region and culture of the world is unique, though interdependent, and is the subject of much study by itself. Unfortunately, such in-depth consideration is beyond the scope of this chapter.

4.4 Surface design techniques (I) – Dyeing

The application of colour or 'dyeing' can be achieved during various stages of the fabric's production. Thus, depending on the end requirement, the dye may be applied to the fibre, yarn or fabric. Technically, dyes are colorants with varying degrees of affinity for textile fibres and are applied dissolved in a medium – usually water – whereas pigments are insoluble colorants with no affinity for textiles. They are applied in dispersions and with a binder to attach them to the substrate.

4.4.1 Methods of dyeing

The industry employs several methods of dyeing and the choice depends on the quantity and form of the material being dyed, as well as the desired quality and cost consideration. These methods are:

Solution dyeing

In solution dyeing, dyes or pigments are added to the polymer solution or melt of synthetic fibres before they are extruded through the spinneret to form the filament fibres. Since the colouring matter is inherently incorporated into the fibre, the colour is permanent and uniform. However, the process is expensive if not done on a large scale, and is used mostly for colouring fibres such as polypropylene, which are difficult to be coloured by other methods.

Stock dyeing

The dyeing of staple fibres before they have been spun into yarn is called stock dyeing. The process is expensive but has advantages of good colourfastness, penetration and level dyeing. It is used predominantly for dyeing wool fibres.

Yarn dyeing

In yarn dyeing the colour is applied at the yarn stage. One of the oldest methods of dyeing textiles, yarn dyeing is particularly useful for fabrics where the pattern results from using different colour yarns in a planned sequence, for example tweeds, plaids and jacquard designs. *Skein or hank dyeing*, applicable on a small scale, is a method of dyeing yarns in the form of loosely wound coils. On a larger scale, yarns are dyed by a process, called *package dyeing*, where yarns wound into one pound packages on perforated cores, are stacked on hollow posts in a dyeing container. The dye solution is then forced upward through the posts and outwards through the perforated cores, thus dyeing the packages inside-out. In *beam dyeing*, the entire warp sheet, wound onto a perforated warp beam, is dyed. In *space dyeing* of yarns, a multicolour effect is achieved by tie-dyeing a skein in multiple colours or printing on the warp sheet or by a knit-print-deknit process.

Piece dyeing

The dyeing of fabric after it has been constructed is called piece dyeing. This is the most widely used, rapid and economical method of dyeing textiles. In *winch* or *beck dyeing*, fabric, usually in a rope form, circulates continuously through the stationary dye liquor, whereas in *jig dyeing*, an open width fabric is passed from one batch roller through the stationary dye liquor to another batch roller and back. The number of to and fro passages through the dye bath determines the depth of colour. Jig dyeing allows bigger batches of fabric to be dyed than beck dyeing. In *padding*, the fabric in open width is passed through the dye liquor or paste and then through a nip which squeezes out the excess liquid in a controlled way. Some special methods for dyeing synthetics are *pressure jet dyeing*, which employs the high temperatures and pressures required to set disperse dyes into the fabric, and *thermosoling*, where infra-red heat is used to dry the fabric padded with the dye liquor, followed by high temperature curing. *Foam dyeing* is a specialty technique used for dyeing carpets, where the dye, mixed with air and only a minimal amount of water, is applied as a foam and in the presence of a carrier, in a specific pattern, to the moving carpet surface. The use of water is thus much reduced.

Union dyeing

Union dyeing is employed for applying a single colour to a fabric composed of two or more different fibres, each requiring a different class of dye. This is achieved either by a single bath process, using a mixture of the different dye classes in the same bath, or by passing the fabric successively through multiple baths, each containing a single dye specific to one of the fibres.

Cross dyeing

Cross dyeing is used to obtain a multicolour effect in a fabric composed of two or more different fibres, by immersing the fabric in a bath containing a mixture of different dye classes in different colours, each colour meant to dye a specific fibre component. In '*cationic dyeing*,' a fabric blend of two synthetic fibres belonging to the same genre, but with varying degrees of dye affinity, is used to create a multicolour effect by dyeing in a single dye bath.

4.4.2 Natural dyes

Natural dyes and pigments are colouring matters of plant or animal origin, and require little or no chemical processing for their application. Most natural dyes require an agent or 'mordant,' which is basically a metallic salt, to help them bind onto the textile substrate. '*Purple*', *kermes*, and *cochineal* are examples of dyes derived from animal origin. The rare and ancient purple dye extracted from the glandular secretion of a shellfish has a very good colourfastness and yields a range of shades from purple to a very dark red. The kermes dye, derived from the dried bodies of the kermes shield louse, and the cochineal dye, obtained from the dried cochineal beetle, give bright red colours. *Woad*, *indigo*, *madder* and *logwood* are some examples of natural dyes of plant origin. Woad, prepared by fermenting the fleshy leaves of the plant *Isatis Tinctoria*, gives a rich blue colour. Indigo, another source of blue colour, is derived by fermenting the leaves of the sub-tropical indigo plant that contains the chemical colouring compound 'indigotin'. The red madder dye, prepared from the dried roots of the plant *Rubia tinctorum*, yields a clear red colour in the presence of the mordant alum. *Lichens*, and *Logwood* are other important source of vegetable dyes (Storey, 1978). The use of natural dyes is almost negligible in modern industry, though it is becoming increasingly popular among small scale suppliers of eco-friendly products.

4.4.3 Synthetic dyes

The production of synthetic dyes started from the mid-nineteenth century with the development of aniline dyes. Since then, the textile industry has rapidly and entirely shifted to the use of synthetic dyes, with a few exceptions. Synthetic dyes are classified into the following groups depending on the method of application and chemical structure. Table 4.1 provides a comparison of the various dye classes.

Acid dyes

Acid dyes are water-soluble dyes used for dyeing protein (e.g. wool and silk), polyamide (nylon) and acrylic fibres. These dyes are applied from an acidic medium and yield very bright colours with varying fastness properties.

Table 4.1 Comparison of various dye-classes

Dye class	Applicable fibre substrate	Properties	Dyeing mechanism
Acid dyes	Wool, silk and other protein fibres; polyamides (nylon), acrylics.	Brilliant shades, variable fastness properties.	Acidic medium required, anions from dyes links to amino groups of protein fibres.
Basic dyes	Acrylics, modacrylics, protein fibres, polyamides.	Bright colours, poor wash and light fastness.	Neutral to alkaline medium required, cations from dyes links to carboxyl group in fibres.
Azoic dyes	Mostly cotton, also rayon, nylon, polyester.	Good light, wash, bleach and alkali fastness; poor crocking fastness.	Colourless diazo compound reacts with napthol to produce colours directly on the fibre.
Mordant dyes	Protein fibres, polyester, rayon, acrylic.	Good fastness to light, washing and dry-cleaning.	A mordant combines with the dye and the fibre, fixing the dye to the fibre.
Direct dyes	Cellulosic fibres (cotton, linen, rayon etc.)	Easily applicable, Poor wash and light fastness without aftertreatment.	Requires a salt to improve dye absorption. They diffuse into fibers and attach through secondary bonds.
Vat dyes	Cotton, rayon, polyester.	Good fastness properties.	Insoluble dyes, converted to an alkali soluble form, applied to fabric and oxidized back to the insoluble form.
Disperse dyes	Polyester, nylon, acetates, triacetates, acrylic.	Good fastness to light, washing, dry-cleaning, fades on exposure to fumes.	Sublimes to vapour state under high temperature and gets absorbed into fibre.
Reactive dyes	Cotton, rayon, wool, silk, nylon, acrylics.	Good wash and light fastness.	Reacts with fibre, forming covalent bonds, needs to be cured with steam or dry heat.

Source: Compiled from Timar-Balazsy, 1998; Wells, 1997; Jackman, 1990; and Storey, 1978.

Basic dyes

Basic dyes can produce bright colours on protein and acrylic fibres. However, because of poor wash and light fastness properties, their use is limited nowadays, except for dyeing acrylic and modacrylic fibres.

Azoic or naphthol dyes

Azoic or naphthol dyes are also called ice colours (since they are applied cold) and are used mostly on cellulosic fibres. When the fabric is treated in napthol solution followed by a colourless diazo compound, the two components react to produce the colour directly on the fibre. Azoic dyes have good light, wash, bleach and alkali fastness, but poor fastness to crocking (transfer of dye from one dry fabric to another).

Mordant dyes

Mordant dyes, which also include premetalized dyes, have no fibre affinity unless they are used with a metallic compound (the mordant) which combines with both the dye and the fibre, thus fixing the dye. Chrome mordant dyes, used for dyeing wool, belong to this class and have good fastness to light, washing and dry-cleaning.

Direct dyes

Direct dyes are water-soluble dyes, used mostly for dyeing cellulosic fibres, and to a limited extent, protein and polyamide fibres. They are easy to apply but have poor wash and light fastness unless aftertreated with resin or chemical developers. Usually a salt, such as sodium chloride (or acetic acid while dyeing protein fibres), has to be added to the dye solution to improve dye uptake.

Vat dyes

Vat dyes are used mostly for dyeing cotton have very good fastness properties. These water insoluble dyes are first converted to a 'leuco' or alkali soluble form by reducing them with an alkaline solution of sodium hydroxide and sodium hydrosulphite. The dye in the soluble form is applied to the fabric, where it gets aerially oxidized back to its insoluble form, thus becoming permanently fixed on to the fabric. Indigo was the first synthesized vat dye.

Disperse dyes

Disperse dyes are used for dyeing synthetic fibres such as acetates, polyester, nylon and acrylic. At high temperatures, these dyes sublime into the vapour state and become absorbed into the fibre, thus enabling dyeing without the use of water. Disperse dyes exhibit good colour fastness to washing, dry cleaning and light, but fade on exposure to gaseous impurities in the environment.

Reactive dyes

The dyeing mechanism of reactive dyes involves actual chemical bonding between the dye molecule and the fibre, thus producing excellent wash and light fastness. Such dyes require a salt (sodium chloride or sodium sulphate) to reduce the solubility of the dye molecules in water and improve their chances of binding with the fibre, and an alkali such as sodium bicarbonate to help bind the dye molecules to the fibre. After the dye is applied to the fabric, it needs to be fixed by dry heat or steaming, and finally rinsed thoroughly to wash off any loose dye.

4.5 Surface design techniques (II) – Printing

Printing is the application of pattern on the fabric surface with dyes, pigments or chemicals by adding, removing, or resisting colours to or from the specified areas of the design, thus creating a contrasting effect with the background. Based on the principle of creating the pattern, there are four main styles of printing – (i) dyed; (ii) direct; (iii) resist; and (iv) discharge. The following sub-sections discuss each of these styles and the techniques employed for their execution. Table 4.2 compares the different printing techniques.

4.5.1 Dyed style

Dyed style is a two-step process of printing. First, the mordant (a metallic salt or oxide that fixes the dye to the fibre) is printed or painted on the fabric in the design areas and then the fabric is immersed in the dye liquor, whereby the colour is taken up only in the areas where the mordant has been applied. The dyed style, owing to the disadvantage of an additional step, has been replaced in the industry by the direct styles of printing.

4.5.2 Direct style

Direct style is the most widely used method of printing in the industry and involves application of the printing paste directly onto the fabric in a

Table 4.2 Comparison of some popular printing techniques

Printing technique	Printing style	Printing method	Comments
Block printing	Direct	Pattern is stamped on fabric by blocks of wood, linoleum, etc, engraved with design in relief.	Traditional hand technique.
Flat-bed screen printing	Direct	Printing paste is forced out by a squeezee through the open design areas of a flat meshy framed screen onto the fabric placed underneath.	Used both in hand operated and automated versions.
Automated roller screen printing	Direct	Same principle as flat bed screen printing, using perforated nickel plated cylindrical screens.	The most widely used printing technique in the industry.
Automated roller printing	Direct	Print paste applied to rotating copper rollers, etched with the designs, are imprinted as patterns on fabric surface.	Expensive process – limited use in the industry.
Heat transfer/ sublistatic printing	Direct	Design printed in disperse dyes on waxed paper is transfered to fabric under heat and pressure.	Fast and efficient process, low dye penetration.
Tie-dye	Resist	Fabric is tied, knotted, folded, clamped, pole-wrapped or stitched and then immersed in dye bath.	Ancient technique, also known as shibori, plangi, tritik, bandhni, etc.
Batik	Resist	Design is stamped, painted or poured in molten wax on fabric which is immersed in dye bath after wax dries.	Ancient technique, also called trap or tjanting.
Ikat	Resist	Pattern is printed on warp sheet and/ or weft yarn before weaving.	Traditional technique, warp printing also done by industry.
Chemical resist printing	Resist	Fabric is printed or painted with a chemical that resists dye fixation then immersed in a dye bath.	Applied mostly on cellulose using pigment-reactive or reactive-reactive resist techniques.
Discharge printing	Discharge	Pattern created by removing colour from a pre-dyed fabric using a reducing agent.	In white discharge printing, colour is discharged to a white colour. In colour discharge printing the discharged colour is replaced by another colour.

Source: Compiled from Nielson, 2007; Wells, 1997; and Jackman, 1990.

single step, followed by fixation. A generic recipe for the print-paste will include the colouring matter, which can be a dye or a pigment, a solvent for the dye or an emulsion for the pigment, the dye auxiliaries, i.e. fixing or levelling agents, and a thickener (such as alginate) to control the consistency of the paste. There are several methods used for printing in the direct style.

Block printing

Block printing is a hand printing technique that has been traditionally used in many cultures around the world. Blocks of rubber, wood or linoleum are prepared by chiselling away the background of the design, leaving only the pattern in relief. Lino blocks, being lighter and cheaper, are sometimes preferred to wooden blocks. Either a brush or a roller is used to apply the printing paste evenly on the raised surface of the block or it is pressed onto a pad soaked in the dye paste. The block is then stamped firmly onto the fabric, stretched and attached on to a flat surface. Each colour of a multicolour design would require a separate block. To ensure proper pattern registration, small metal pins inserted at the corners of the blocks act as guides for placing the blocks.

Screen printing

Screen printing is a versatile method of printing and consists of a screen of sheer silk (organdie), nylon or polyester stretched over a wooden or metal frame. The screen, converted into a stencil by filling out its mesh in the negative parts of the design, is firmly positioned on the fabric and the printing paste is forced through it, onto the fabric placed underneath, by the dragging action of a rubber blade called a squeegee. This process is called *flat bed screen printing*. For multicolour designs, as many screens as the number of colours are required. In industry, the screen used is a photographic stencil, made by exposing a light sensitive screen, blocked by the positive transparency of the design, to light. The exposed negative parts of the design harden under the effect of the light, while the remaining areas stay soft and are eventually washed away. The photostencil allows the very accurate colour separation required for printing multicolour designs. In *automated flat-bed screen printing*, the entire process is automated and the screen is lifted and lowered in the same place while the fabric moves forward on a conveyor belt. *The Zimmer flat bed screen printer*, designed specially for carpets, uses magnets or vacuum suction beneath the carpet to pull the dye paste downwards. *The automated rotary screen printing* machine uses a series of perforated nickel plated cylindrical screens – each cylinder for one colour of the design. The screens are stencilled with water-soluble lacquer. The machine allows upto 20 cylinders, accurately printing different coloured areas on the fabric moving on a conveyor belt.

Roller printing

In roller printing, a copper printing roller with the design engraved on its surface and fed with the dye paste from another roller, imprints the design on the fabric that is rolled against its surface under pressure by a larger cylinder. Each printing roller is provided with a doctor blade to scrape off any excess dye or lint picked up by its surface. There are as many printing rollers as the number of colours in the design and their precise placement and rotation around the fabric roller ensures the correct registration of each colour in the design. Roller printing, being an expensive process, has been completely outdated by rotary screen printing.

Heat transfer or sublistatic printing

Heat transfer or sublistatic printing employs high temperature and pressure to transfer a design printed in disperse dyes on a waxed paper, onto a fabric surface. The heat and pressure causes the dye to vaporize from the paper and sublime onto the fibre, with which it has a greater affinity. Transfer printing, being fast and efficient, allows flexibility in production but has the disadvantage of low dye penetration.

4.5.3 Resist style

In this style of printing, the dye is prevented from reaching certain areas of the fabric by a physical or chemical barrier. A physical barrier may be a tied knot as in 'shibori', or a layer of wax as in batik printing. A chemical barrier might be an organic acid used to prevent fixation of reactive dyes onto cellulose. Most methods of resist dyeing using physical barriers are very old traditional techniques. Some of the popular resist styles of printing are:

Tie-dye

Tie-dye is an ancient patterning technique, used in many cultures and known by different names, such as 'shibori' in Japan, 'plangi' and 'tritik' in Indonesia and Malaysia and 'badhni' in India, to mention a few. The basic process involves binding, knotting or tying bunches of the fabric before immersing it in the dye bath. Alternatively, the fabric can be stitched, folded, clamped or wrapped around a pole and gathered before applying the dye. The resulting effect in each case is a pattern of undyed areas where the dye could not reach the fabric. This process can be repeated to produce a layered effect.

Batik

Also known as tjanting or trap, batik is a very old technique of resist dyeing. In the process, molten wax (usually a mixture of paraffin and beeswax) is applied to the fabric by painting, stamping, or pouring from a special container called 'tjanting'. When the wax dries, the fabric is immersed in the dye bath. After dyeing, the removal of the wax from the fabric leaves undyed areas, where the wax resisted the dye, in sharp contrast with the dyed background.

Ikat

Ikat is a special effect, created in a fabric woven with a warp sheet and/or weft yarns that has been tie-dyed prior to weaving. This process requires a lot of precision and skill. The resultant effect after weaving is a beautiful defocused pattern.

Chemical resists

Chemical resist is employed mainly for cellulosic fibres and involves the use of a chemical, which may be an acid, alkali, salt, reducing or oxidizing agent, to prevent the development or fixation of the colour on the fabric. Pigment-reactive and reactive-reactive are the most widely used chemical resist technique.

4.5.4 Discharge style

In this style, the pattern is created in negative by the removal of the ground colour from a pre-dyed fabric with a reducing agent such as sodium formaldehyde sulphoxylate or zinc formaldehyde sulphoxylate. Common methods of applying discharging pastes are screen printing (most widely used), stamping, stencilling and hand-painting. In *white discharge* printing, the fabric is discharged to a white colour. In *colour discharge* printing, undischargeable dyes are added as illuminating colours in the discharge paste. The colour discharged from the ground is replaced simultaneously by the illuminating colour.

4.6 Surface design techniques (III) – Creating textures

A variety of textures can be created on the fabric surface by physical or chemical means, causing the fabric to pucker, crease, shrink, burn out, shine, fuzz up, or assume a multitude of three dimensional shapes. The following subsections deal with some interesting textural effects that can be created on fabrics in an industrial or studio set up.

4.6.1 Crimping

Crimping or puckering of the fabric can be caused by a chemical such as caustic soda or by the application of a rubber solution or puff pigment. A *plisse* is created by screen printing or hand-painting a pattern of strong caustic soda solution on a cellulose fabric. The chemical shrinks the fabric in the printed areas, causing the neighbouring regions to pucker. The effect can also be produced by resist printing a thick solution of gum arabic on the fabric before immersing it in a caustic soda bath. In *puff pigment printing*, the pigment is screen printed onto the back of a fabric such as lycra or velvet, stretched tightly, followed by heat fixing once the binder has dried. The heat causes the pigment to puff up, creating the pucker. *Latex* or *synthetic rubber* solutions can also be screen printed in a similar way, but without heat-setting, to create puckering in stretchy fabrics.

4.6.2 Embossing

Embossing is creating a pattern in relief. This is achieved by passing the fabric between the heated surface of a metal bowl engraved with the design, and the soft surface of another bowl pressed against the former. This heat-set effect can be created on synthetics or on wool treated with a resin. *Moiré* is also an embossed effect of watermark lines created on a fabric that has fine horizontal ribs, by a calendering roller engraved with a wavy pattern.

4.6.3 Applying gloss

Glossy finishes can be achieved in several ways. *Cire* or *chintz* is a glossy finish applied on thermoplastic fibres, e.g. nylon or polyester, by calendering them under heat and pressure and on silk or rayon by calendering them along with a wax or resin application. *Panne* finish is a luster-enhancing treatment given to pile fabrics such as velvet by passing them through heavy rollers to flatten down their pile in one direction.

4.6.4 Flocking

Flocking, or creating a velvet-like finish, is achieved by screen printing adhesive and then spraying flocks, or tiny cut fibres, onto the fabric surface. The fabric is vacuumed, cured and cleaned to complete the finishing. In *electrostatic flocking*, an electric charge is used to make the flocks stand up on their ends during the treatment, thus enabling better alignment and a cleaner finish.

4.6.5 Napped finish

A fuzzy or napped finish is imparted by raising the fibre ends on the surface of a fabric, either by brushing with rollers covered in wire bristles (*napping*) or by abrading with rollers having a surface similar to sand paper (*sueding*).

4.6.6 Burn-out

A burn-out or etched effect can be created by partial destruction of a fabric with an etching agent such as sulphuric acid. Fabrics suitable for this process should have two fibre components – one susceptible to the etching agent and the other resistant. Thus, when sulphuric acid is printed on a fabric composed of polyester and cotton, the acid destroys the cellulose component of the fabric in the printed areas, creating a translucent, meshy pattern. *Devore* is a popular burn-out velvet, widely used in interiors, in which the pattern is created by destroying the pile.

4.7 Surface design techniques (IV) – Embellishing fabrics

In addition to the multitude of construction and finishing techniques used, the fabric can also be worked upon or embellished in a variety of ways to enhance its surface. Most embellishing methods, such as embroidery, quilting, appliqué and beadwork are age-old techniques of decorating fabric and are still largely executed by hand. A specialized industrial sector in trimmings, lacework, embroidery and other embellishments has also developed that uses innovative technology to recreate the hand-worked effects on an industrial scale and pace. A few embellishment techniques are briefly discussed below.

4.7.1 Embroidery

Decorating fabric with stitches in yarn or thread, using a needle, is one of the oldest forms of art. While the library of embroidery includes hundreds of varieties of stitches, they can be categorized broadly into four main techniques – *raised work* or stumpwork where raised effects are created by stitching over pads of wool and cotton; *couched work* – creating a pattern with cords by sewing them onto the base fabric; *flat running* and *filling stitches* of which there are hundreds; and *counted thread embroidery*, e.g. needlepoint and cross stitch, where the stitches are placed over a counted number of threads of the base fabric. *Schiffli embroidery* is an example of machine embroidery made on the 'Schiffli' machine that works sideways with a thousand needles. This machine embroiders with a top, decorative and a back, binding yarn, and is used for making laces and sheer curtain fabrics. Many embroidery stitches can now be produced in *digitized embroidery machines* which, for large scale production, have multiple heads to produce a number of identical designs simultaneously using an embroidery software program.

4.7.2 Quilting

Quilting is the technique of stitching together, by hand or machine, multiple layers of fabric with a filling of cotton, foam or polyester batting in between the layers. Quilting is widely used for making bedspreads, quilts, comforters etc. Single-needle, hand-guided quilting machines are used for making *outline quilting*, where the stitching lines follow the outlines of the print design, *vermicelli*, which uses free motion all-over stitching patterns, and *trapunto* or *Italian quilting*, a form of 'couching' where a cord inserted and stitched between the fabric layers creates a raised pattern. In mass-scale automated production, multiple-needle machine quilting is used to make simple geometric patterns. In *stitchless quilting*, multiple layers of fabric are fused together thermally or by an adhesive, creating the appearance of being stitched.

4.7.3 Appliqué

In appliqué, small pieces of fabric or other material are couched or stitched onto a base fabric. In reverse appliqué, the base fabric is on top of the stitched fabric and the top fabric is cut out to reveal the appliquéd fabric underneath.

4.7.4 Patchwork

Patchwork, used mostly for making bed quilts and cushions, is the technique of creating a fabric layer by joining small pieces of fabric (traditionally scraps of old clothes or textiles) in geometric or abstract patterns. Being a hand-worked technique, it is mostly produced on a small scale.

4.8 Technological advancement

Innovation in surface design techniques of textiles, like in all other production processes, has been in multiple directions. Environmental pollution has been a harmful side-effect of most of the dyeing, printing and surface treatment finishes developed in the twentieth century and beyond. A large number of new products and processes are geared towards alternative, eco-friendly solutions of surface finishing processes.

Textile design and printing technology have greatly benefited since the advent of the digital age in the late twentieth century. *Computer aided textile design*, digital printing and digital embroidery are some of the textile processes that can now be executed with a greater choice and control than ever possible before. According to the report, 'Developments in textile colourants' published by *Textile Outlook International* in their 127th issue (2007), *Digital printing* of textiles using ink jet printers constitutes only 1% of all industrial printing at present. But it is getting increasingly popular and is expected to have a 10% segment by the beginning of the next decade. Improvements in the printhead, ink dispersion techniques (such as the valve-jet process that uses valves which dispense ink droplets at high velocity and pressure, and are capable of opening and shutting independently) and the availability of an increasing choice of dye and pigment based ink-jet inks specifically for textiles, are some of the reasons behind the increasing popularity of digital printing. In *digital embroidery*, computer software allows the pattern to be designed with a wide choice of stitch parameters, and the machine executes the embroidery entirely on its own.

The use of *nano-technology* (the science involved in the development and control of materials and devices in the size range of 100 nanometers and less), in imparting functional enhancement finishes on textiles, is another area that has seen many innovations in recent years. An interesting example of the application of nanotechnology in the surface design of textiles, is a polyester fabric created by Teijin Fibres in Japan. The fabric acquires a mystical hue by applying on a polyester substrate about 60 layers of polyester and nylon of about 69 nanometer thickness, each of which refracts light with a different refractive index.

Electronic textiles are a range of 'smart' or 'intelligent' textile products developed in the 1990s. In these textiles, functionality of the fabric is enhanced by incorporating electronic or computing components unobtrusively into the fabric. Depending on the kind of electronics integrated, the resulting fabric can be used, for example, as platforms for communicating, sensing and interacting with the environment.

4.9 The role of designers, trend forecasters and trade fairs

4.9.1 The designer's contribution

In today's competitive global businesses, it is imperative for the home textile industry, as any other, to be in 'sync' with the current seasonal trends. Design teams require information of future trends at least a year in advance to allow sufficient time for planning and production, before a range of products can be brought to market at the right time. Thus, the designer has to remain constantly educated in long and short term trend forecasts in colour, pattern, texture, new fabric, new yarns, innovative materials and anything else that can be a useful input in the design process. This is achieved through regular visits to trade fairs and trend seminars, inviting trend specialists to present colour and style forecasts for each upcoming seasons, and a constant updating of design databases. Usually, every company targets or emphasizes a specific segment of the entire spectrum of the market. The role of the designer working for the industry is to interpret information received from forecasting agencies and to implement them in a way that works best for the specific product and price range of the company.

4.9.2 Trends: Their source and importance

An organization can develop its unique marketing strategy or technology to get an edge in the industry, but when it comes to choosing colour and style, it has to follow the general direction of the market – the direction that has been set into motion by highly specialized individuals or groups of trend professionals. Most companies rely on design consultants and prediction agencies for their constant source of information on trend forecasts.

The forecasting agencies have their team of experts, whose job is to analyze patterns, and changes in consumer tastes, behaviour and culture over time and geographical boundaries. The influence of factors such as new technologies, economy, ecology or politics on the consumer's psychology is studied carefully. A combination of intuition and observation from the studies is used to predict the direction of the consumer's choice in the short term and long term future. These predictions are then presented and sold to the industry in the form of trend-books, seminars, fashion shows, or online resources. While the industry processes the information and translates them into production, the media informs the consumers about what they are supposed to want next season in order to be trendy. Whether the people's choices are the source of trend inspirations or the other way round, the influence of trends and forecasting agencies on the industry, as well as the consumers, has risen unabated through the years.

The general format for trend presentation in trend books and seminars is thematic – there would be a few themes explaining new trends in cultural outlook and choices of the consumer. Associated with each theme would be suggestions of a colour palette (usually referred in Pantone colours), materials, and visual representations of the essence of the theme.

4.9.3 Forecasting agencies

The trend for forecasting colours started around the 1960s with the establishment of organizations such as the International Colour Authority (ICA) and the Color Marketing Group (Gale, 2002). Now there is a plethora of agencies and experts on style and colour forecasts, which work for a wide range of products and industries. The ICA is known to be the earliest colour forecaster and presents its predictions 24 months in advance of the retailing season; while the Color Marketing Group provides trend information about 19 months in advance. Worth Global Style Network, founded in 1998, use the internet to deliver their colour and design forecasts and in no time their online resources have become one of the most popularly subscribed and sought after source of news and information on trends (Gale, 2002). The Centre for Promotion of Imports for Developing Countries, located in the Netherlands (abbreviated CBI in Dutch) provides free information on fashion and colour trends, along with a range of guidance on health and safety standards and legislations to the industries in the developing world (Anon., 2001).

As for individual trend experts, American forecasting pundit *Faith Popcorn* and the Dutch forecaster *Lidewij Edelkoort* stand out for their influence and popularity (Gale, 2002). Faith Popcorn's trend consultancy service, 'BrainReserve'; offers advice on marketing strategies and future trends to a wide range of clients including Walmart, Starbucks, BMW and MySpace. Edelkoort's agency 'Trend Union' consults especially for the Fashion, textiles and interiors industry.

4.9.4 Trade fairs

Trade fairs are increasingly becoming platforms of business and information exchange for manufacturers, buyers, agents, distributors, and design professionals in the global world of interior textiles. The fairs serve not only as events of product exhibition and sales contact between manufacturers and buyers visiting from around the world, but also as forums for exchange of ideas and concepts. In fact, almost all textile trade fairs incorporate a section where new materials, future concepts, and trends are exhibited by companies and designers.

The 'Heimtextil', hosted in Frankfurt, Germany, is one of the most important annual trade fairs for interior residential textiles mainly. 'Decosit', a residential textiles fair along with its contract textiles counterpart 'Decocontract', organized in Brussels, is another significant fair for interior fabrics and furnishing. Decosit also has a section called 'Decotec', dedicated to the display of a library of innovative and conceptual materials for outdoors and indoors, and 'Indigo', which is an exhibition of designs from textile design studios around the world. Other important textile fairs include 'Premier Vision', Paris, primarily for fashion fabrics; 'Expofil' for innovative fibres and yarns; 'Interstoff' in Frankfurt, Germany; 'Pitti Filati' in Florence, Italy, for textile yarns; and 'Top Drawer' in London, for interior and fashion accessories.

4.10 The future of design for interior textiles

4.10.1 The environment and society

The problems and dilemmas of attaining ecologically and socially sound standards of production are manifold and universal. Questions such as how to achieve environmentally safe manufacturing while maintaining economic sustainability, how to survive in a cut-throat global competition with an ethically fair production chain or how to tackle issues of wastage with the pressures of keeping up with trends, have no straightforward solutions. The textile sector in particular has been one of the biggest perpetrators of environmental pollution and socially unethical manufacturing, and the consumer has never been more aware about it than they are now.

The growing demand for eco-friendly products from an ever-increasing segment of the consumer population, whether out of genuine awareness or as a result of politicized propaganda, has kicked off a booming business in 'green products'. With a comprehensive solution from the industry nowhere near in sight, this trend is being predicted to stay and influence the industry for quite some time to come.

4.10.2 Eco-friendly textiles for the interior

Environmental damage caused by the textile industry can be considered to occur in two stages (Roy Chowdhury, 2006):

(i) At the raw material and manufacturing stage. Notable examples would include the use of large quantities of pesticides in cotton cultivation, the presence of harmful adsorbable organic halogen compounds (AOX) in agents for bleaching, shrink-proofing, and insect-proofing of wool, and the use of harmful formaldehyde-based resins for improving wrinkle resistance of fabric.

(ii) At the post manufacturing stage. This category includes problems such as waste disposal and the generation of huge quantities of non bio-degradable wastes.

Pressures faced by the textile and fashion industry to rectify the problems of ecologically non-sustainable production come from legislation and restrictive trade policies, and also, to an increasing extent, from consumers. The industry has responded in varying degrees by adopting corporate environmental policies to develop a steady path for improvement, acquiring information and organizing training and workshops and obtaining certification from bodies such as GOTS (Global Organic Textile Standards), Skal (A Dutch certification program), and Oeko-Tex (standards established by an association of textile testing institutes situated worldwide). Textile manufacturing companies are also investing in research and development of innovative eco-friendly products and processes. A few examples in this endeavour, especially within the context of surface finishes and design are:

AirDye technology

AirDye technology was developed by Transprint USA, now a subsidiary of Colorep Inc., and uses air instead of water as a medium of dyeing, resulting in a 70 percent reduction in energy requirement, as compared to conventional cationic and vat-dyeing processes. Both the home furnishings and apparel industries have been targeted as prospective users of this technology (McAllister, 2008).

Bio-degradable sulphur dyes

Two new sulphur dyes, 'Diresul Indinavy RDT-B liq.' and 'Diresul Indiblack RDT-2R liq.' designed for dyeing cellulosics, are claimed to be 100 perecnt bio-degradable have been launched by the Swiss-based company Clariant. These dyes and free of any amines, heavy metals or halogens (Anon., 2008).

High-end eco-textiles

Interior designer Emily Todhunter has collaborated with the company 'O Ecotextiles', known for their environmentally-friendly natural fabrics, to produce a range of luxury interior fabrics that meets very stringent standards of ecologically sound processing (Alexander, 2007). Her collections use linen, hemp, organic cotton, and ramie fibers, processed in an ecologically conscious manner using organic dyes and oxygen-based bleaching. The collections, presented in 'Decorex' 2007, have been acclaimed as some of the pioneers in high-end eco-textiles in interiors.

Recycled polyester

'Repreve' is a 100 percent recycled polyester yarn manufactured from 20 percent post consumer and 80 percent post industrial polyester, by the textured yarn company Unifi Inc. 'Repreve' has been used specially in the contract interior textile industry in draperies, upholstery, and wall covering (Rudie, 2006). The use of recycled polyester saves a lot of energy and water that are usually consumed in conventional yarn processing, and also reduces the problems of landfill.

Fortunately, the list of successful implementation of capital by the textile industry towards achieving more ecologically sustainable production is growing.

4.10.3 Collaborative research and unified trend directions

The commonality of issues such as ecological sustainability and ethical responsibility that have been of growing concern in the design process of a wide variety of products, has encouraged much collaborative work between designers and researchers belonging to various areas in the industry and academia alike. This has resulted in an increasing amount of lateral thinking and interdisciplinary work between various sectors of the industry and also between academia and the industry.

Another factor which unifies the design directions of different industries is the market trends. The basic long-term trends proposed by the forecasters are, to a large extent, common within different areas of a sector or even between different sectors. The boom of 'green products' is an example of shared vision between different industries, based on a common trend.

4.10.4 Effect of technological innovations

Apart from changes in consumer tastes, technological innovations also have an important influence on the future directions of industries. A few of the popular and recently developed technologies that have been widely adopted by a diverse range of industries, including textiles, are nanotechnology, biotechnology, and the technology of 'smart materials'. All of these technologies have had far reaching effects on the future of design and

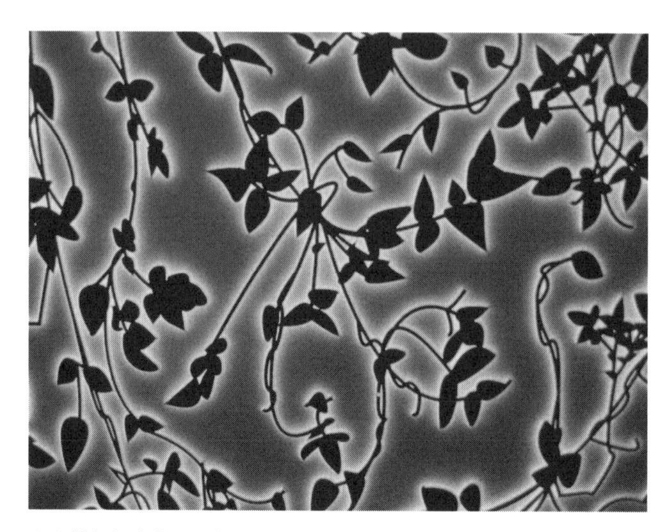

4.1 'Digital Dawn' (created by Studio Loop.pH), is an interactive window blind which uses electroluminescence and incorporated light sensors to change its light intensity in response to the surrounding light.
Source: http://www.loop.ph/bin/view/Loop/LoopPressImages. Image copyright to Studio Loop.pH.

innovation of interior textiles. Nanotechnology has been successfully used to develop a range of textile finishes such as soil, water and dirt repellent finishes, antimicrobial finishes, and fragrance and biocide releasing finishes, to name a few. Similarly, biotechnology has been widely applied, through the use of enzymes in replacing environmentally harmful chemicals, for many finishing processes of textiles. 'Smart textiles' or 'electronic textiles', include a range of electronics incorporated into textile products. Some beautiful examples of surface design that has been created through smart textiles are the interior products of 'Loop.ph', a London-based research and design studio (website: http://www.loop.ph/bin/view/Loop/WebHome). These include interactive screens and blinds with light emitting patterns that respond to the environment and are created with electroluminescent technology. Figures 4.1 and 4.2 show images of a couple of their products. Design research initiatives such as these are the pioneers of interactive and thought provoking textile design for the future – designs based on a convergence of aesthetics, science and technology.

To sum up, the future of successful design for tomorrow's interiors demands a dedication for knowledge, responsibility, and openness of mind – knowledge about the latest in technology and trends, responsibility towards society and the environment, and an openness to reach beyond all barriers of conventional thinking.

4.2 The 'Blumen Wallpaper', another creation of Studio Loop.pH that uses printed electroluminescent patterns on sliding panels which glow in response to the environment.
Source: http://www.loop.ph/bin/view/Loop/LoopPressImages. Image copyright to Studio Loop.pH.

4.11 Sources of further information and advice

For information on new materials, 'Materio' is a great resource to consult. This online database on new materials can be accessed on subscription through their website http://www.materio.com/. For inspirational ideas, the website http://www.tfrg.org.uk/mission, showcasing projects of the 'Textiles Futures Research Group' – an organization that promotes researches interfacing textile design with science, technology, culture and ecology – is worth visiting. 'Textile View', 'Viewpoint' and 'Bloom' – all publications of Trend Union – are good inspirational magazines for surface design ideas. For information on trend news and reports, the website of World Global Style Network (http://www.wgsn.com/) is a comprehensive one-stop shop. Better still, is their subscription-free student website http://www.wgsn-edu.com/edu/. For textile trade shows and events, the website http://www.biztradeshows.com/textiles-fabrics/ provides an exhaustive list of all upcoming trade fairs and expos in textiles, throughout the world.

4.12 References

ALEXANDER, M. C., 2007. Eco home. *Telegraph*, [internet] 24 November. Available at: http://www.telegraph.co.uk/property/main.jhtml?xml=/property/2007/11/24/pecohome124.xml [Accessed: 7 September 2008].

ANON, 2001. Don't be off-colour. *International Trade Forum*. [internet]. Available at: http://www.tradeforum.org/news/fullstory.php/aid/418/Don%92t_Be_Off-colour.html [Accessed: 4 September, 2008].

ANON., 2008. Clariant launches new sulphur dyes. *Ecotextile News*. [internet] 11 September. Available at: http://www.ecotextile.com/news_details.php?id=860 [Accessed: 14 September].

GALE, C. and KAUR, J., 2002. *The textile book*. Oxford: Berg.

JACKMAN, D. R. and DIXON, M. K., 1990. *The guide to textiles: For interior designers*. 2nd ed. Winnipeg: Peguis.

LEBEAU, C. DIRAND, J. and CORBETT, P., 1994. *Fabrics: The decorative art of textiles*. New York, NY: Potter.

MCALLISTER, R., 2008. USA: Technology helps manufacturers become more green. *ifashion* [internet] 30 June. Available at: http://www.ifashion.co.za/index.php?option=com_content&task=view&id=921&Itemid=201 [Accessed: 7 September 2008].

NIELSON, J. K., 2007. *Interior textiles: Fabrics, applications, and historic styles*. Hoboken, NJ: John Wiley & Sons Inc.

ROY CHOWDHURY, A. K., 2006. *Textile preparation and dyeing*. Enfield, NH: Science Publishers.

RUDIE, R., 2006. Great Chemistry. *LDB interior textiles*. [internet] July. Available at: http://www.ldbinteriortextiles.com/pdf/July2006/GreatChemistry.cfm [Accessed: 8 September 2008].

STOREY, J., 1978. *The Thames and Hudson manual of dyes and fabrics*. London: Thames and Hudson.

TIMAR-BALAZSY, A. and EASTOP, D., 1998. *Chemical principles of textile conservation*. Woburn, MA: Butterworth-Heinemann.

WELLS, K., 1997. *Fabric dyeing and printing*. Loveland, CO: Interweave Press.

4.13 Bibliography

ANON., 2007. Digital textile printing on growth trajectory. *Just Style*. [internet] 20 April. Available at: http://www.just-style.com/article.aspx?id=97071 [Accessed: 15 September 2008].

EDWARDS, C., 2007. *Encyclopedia of furnishing textiles, floorcoverings and home furnishing practices, 1200–1950*. Aldershot: Lund Humphries.

GREEN, D., 1972. *Fabric printing and dyeing*. Newton Centre, MA: Charles T. Branford.

LARSEN, J. L. and WEEKS, J., 1975. *Fabrics for interiors: A guide for architects, designers, and consumers*. New York, NY: Van Nostrand Reinhold Company.

PROCTOR, R. M. and LEW, J. F., 1984. *Surface design for fabric*. Seattle, WA: University of Washington Press.

5
The use of textiles in carpets and floor coverings

D. WHITEFOOT, The Carpet Foundation, UK

Abstract: This chapter traces the history of carpet manufacture from historical times up until the present day. It also includes detailed information on carpet properties and the yarns and fibres used as raw materials. There is a detailed comparison of manufacturing methods used and the most recent developments are identified. A brief note is also included on possible future trends for carpet manufacture.

Key words: carpet properties, Axminster, Wilton, tufted, non-woven, fibres.

5.1 The role of textiles in floor coverings

The first floor coverings were likely to have been animal skins or vegetable matter, scattered on the floor, and the earliest floor coverings of textile construction were probably crudely woven rushes or grasses. Inevitably, the desire eventually arose to make floor coverings with a pile of sheep or goats wool to utilise the warmth and comfort of animal skins. Inevitably no traces of these floor coverings remain.

The oldest known existing carpet is 2400 years old and is now to be found in the State Hermitage Museum in St Petersburg, having been excavated from a tomb in Southern Siberia. It is a pile rug of fine construction with about 3600 tufts/dm^2.

5.2 Residential use

Carpets for residential use are made throughout the world, with particularly important centres in the USA and Western Europe. Important areas of carpet manufacture are emerging in the Indian sub-continent and in China. The products made in each of these centres have evolved in different ways and display different characteristics of style. In general terms, US carpets often have a pile of polyamide in patterned and textured loop, from polyester in longer cut-pile referred to as 'Saxonies', and more budget-conscious constructions with polypropylene pile.

Western Europe, other than the UK, makes a proliferation of tight, low pile constructions, with polypropylene the dominant fibre and with polyamide for the better end of the market, whilst UK manufacturers retain a more traditional approach with wool remaining the dominant pile fibre, usually in relatively luxurious cut-pile styles. At present, the most characteristic UK carpet style is a hard-twist cut-pile plain carpet with a wool-rich pile fibre blend.

In the UK, which has historically had a large woven carpet capacity, heavily patterned carpet styles have dominated for many years and there has been a desire to develop less labour intensive, higher speed manufacturing methods. Recent trends, influenced by large volumes of less expensive tufted carpets imported largely from Belgium and Holland, and through lifestyle programmes and articles in the media, have seen the former dominance of the patterned carpet disappear in favour of the plainer styles that are preferred for modern living. Recently, the media has suggested that patterned carpets are now making a comeback, but this is not reflected in carpet production and may be a case of the media trying to establish a trend.

5.3 Commercial use

Carpets intended for commercial use are often subjected to greater concentrations of traffic and need to withstand this. Commercial use is varied and can include offices, retail premises, hotels and leisure centres, casinos, theatres and airports. Carpets for offices, particularly modern open-plan offices and for some larger retail premises feature hard-wearing, dense, low loop-pile constructions, usually with polyamide pile. The carpet tile has found particular favour for this end-use.

Carpet tiles, frequently available in 50 cm square or 45.7 cm (18 inch) square formats, lend themselves particularly to large, multi-floor installations. Of particular importance is the ease with which the tiles, packed into easily handled boxes, can be transported to upper floors. The original carpet tiles were of a simple felt face layer bonded to the tile backing. Styles have developed from these through needled pile floor coverings, plain loop-pile tufted, plain cut-pile tufted and even patterned constructions, each stage gaining in style, luxury and sophistication.

The carpet tile is a composite material composed of a textile surface bonded to a tile backing, which must have the basic important properties of dimensional stability and the ability to lie flat on the floor without doming at the centre of the curling up of tile edges. Installation of early tiles gained added interest by adopting what is called chequer-board installation, where the manufacturing direction (always denoted by an arrow printed or embossed on the back of each tile) is laid at right angles to the adjacent tile. This not only changed the surface aspect of the entire installation but also disguised (through emphasis) the tile edges. With more sophisticated and luxurious styles, this simple or utilitarian effect was less desirable and the requirement arose for a uniform appearance, which demanded carefully cut tiles and was more easily achieved with denser carpet constructions and cut pile styles.

Patterned carpet tiles must have either a totally random design or a carefully engineered design repeat to fit exactly into the tile dimension. Simple tufted carpet designs, achieved with stitch displacement devices, appeared initially attractive but the failure of the design repeat dimension, often quite small, to fit exactly into the tile dimension created an effect, colloquially known as 'zippering', to occur at tile edges which may be more or less acceptable to the end user. Sophisticated larger repeat designs, sometimes achieved using a woven carpet surface, or a printed design exactly registered to the tile dimension were inevitably introduced. These also include corporate logos which could be inset into plain or lightly patterned areas.

The ability of the carpet tile to be easily cut in any direction without fraying lent itself to the creation of particularly large logo installations using a number of tiles by carefully creating shapes of different colours and assembling them on the floor.

For the hotel, theatre and airport market the heavily patterned woven carpet remains important on a global scale. The property of heavily patterned carpets to disguise the inevitable effects of concentrated wear and soiling is extremely important for these end uses and the use of wool-rich pile significantly reduces ugly marks resulting from burns and scorches from dropped cigarettes.

5.4 Carpet properties

5.4.1 Aesthetics

Carpets are available in a wide variety of styles, textures, designs and colours, which the skilled interior designer can use to create a stylish interior suited to the activity conducted in the carpeted area. In the home, carpet helps to provide a warm and comfortable environment away from the harsh realities of the everyday world. In commercial buildings, carpet helps to make a statement about the enterprise and creates an environment conducive to efficiency.

5.4.2 Thermal insulation

Carpets are excellent thermal insulators, a property that is enhanced further by a good underlay. With conventional heating systems, the insulation properties of carpet and underlay can significantly reduce heat loss through the floor. Depending upon construction and specification, carpet may have a thermal insulation varying typically between about 0.1 m^2K/W and 0.3 m^2K/W.

In the case of underfloor heating however, the apparently excellent thermal insulation properties of the carpet do not excessively impair the efficiency of the heating system. It is believed that this is a function of the fact that in an underfloor heating situation, the carpet becomes the heat transmitter to the airspace above. However, excessively thick and luxurious carpets, particularly when installed on a thick felt underlay, can be expected to slow the transfer of heat from floor surface to airspace to an unacceptable level.

Within the EU, the thermal insulation properties of carpet are determined according to ISO 8302 *Thermal insulation – Determination of the steady-state thermal resistance and related properties – Guarded hot plate apparatus.* In the UK, the thermal insulation value of carpets has been measured according to BS 4745 (Togmeter) test but this is no longer considered to provide a reliable guide since it has been determined that, compared with actual usage, it 'overstates' the thermal resistance by a considerable margin.

5.4.3 Acoustic insulation

Changes in lifestyle in recent years, in which smooth floors, particularly wood and laminate floor coverings have gained in popularity, has demonstrated to the public just how noisy an uncarpeted room can be. Carpeting is one of the most effective ways of reducing noise and the best carpets can provide acoustic insulation approaching the same level as dedicated acoustic insulation materials.

There are three ways in which carpet can provide acoustic insulation. Possibly the most important of these is impact sound absorption. This is concerned with the way in which sound of, say, a footfall or a dropped object is transmitted into the room below. The pile of a carpet significantly reduces the energy of the impact and has the effect of converting a sharp, high-frequency sound into a low-frequency thud, which has significantly less impact on the ear.

Impact sound absorption is determined according to ISO 717-2 using a device called a tapping box, which drops small hammers onto a floor surface to generate the noise, and measuring the sound generated in a room below, all under controlled standard conditions. Comparison of the generated sound through the bare floor with that through the carpeted floor, as a decibel ratio, can be used to compare floor coverings.

Airborne sound reduction is measured according to the test method described in BS EN 20354 and the Sound Absorption Class may be derived from this according to BS EN ISO 11654. The test method involves a purpose-built reverberation chamber with three loudspeakers, and five microphones per loudspeaker with which the reverberation times at one-third frequency intervals in the range 100 Hz–500 Hz may be measured. The test is conducted with the floor carpeted and the result compared with the figures taken when the floor is bare. The sound reduction co-efficient is calculated taking into account the area of carpet, dimensions of the test chamber, etc., and averages the reduction in reverberation times at each frequency band. A sound reduction co-efficient of 1 would be a perfect insulator and 0 would represent a perfect reflector. The more luxurious carpets will have a sound reduction co-efficient as high as 0.5–0.7, equivalent to acoustic ceiling tiles and sufficient to comply with the UK Building regulations for circulation areas in public buildings.

Carpet, of course, also significantly reduces impact noise in the room in which it is installed.

5.4.4 Safety

The textile surface of a carpet and its three-dimensional structure make carpet a particularly safe surface on which to walk. The carpet surface will have excellent slip resistance and will offer a soft, forgiving surface should falls occur.

From 1st January 2007, carpets sold throughout the EU are required to comply with the health, safety and energy saving requirements of the Construction Products Directive. These are described in detail in EN 14041 which also describes the necessary labelling, required in most EU countries, of the associated CE mark.

5.4.5 Impact upon human health

Until fairly recently, many articles appeared in the popular media which asserted that carpet presented a health hazard, particularly in respect of asthma and other allergic diseases. However, a review of scientific literature, particularly those papers issued in the last few years, suggest that the opposite is, in fact, the case. No scientific evidence has been found that proves that the removal of carpet alone has a clinical benefit since carpet removal has been exercised together with a number of other actions which might alleviate symptoms. The three-dimensional construction of a carpet with pile is such that it can entrap the fine allergen particles which give rise to health problems when inhaled, until removed from the carpet by periodic vacuum cleaning. A carpet has, therefore, an important role to play in significantly reducing the allergen content of indoor air.

Wool, in particular, is known to absorb gaseous toxic pollutants from the atmosphere such as formaldehyde, sulphur dioxide and oxides of nitrogen. The large fibre surface presented by the pile of a carpet allows significant amounts of such pollutants to be absorbed, thereby contributing to improved indoor air quality.

5.5 Types of textiles used as floor coverings

5.5.1 Fibres utilised in the use surface

Wool

Wool remains the most popular pile fibre for carpets made in the UK. Carpet wools are coarse (>33 μm), crimpy and often medulated. Wools from British mountain and moorland sheep breeds are particularly popular but have the disadvantage of containing kemp and dark coloured fibres. For carpets of plain, pale colours, a white or near white wool, free from dark hairs is required and for these styles wools from New Zealand sheep breeds are necessary. New Zealand wools are, in general, considered softer and more lustrous.

Carpet wools often come from sheep bred primarily for meat production and can be considered a by-product, but some New Zealand sheep are bred for the dual purpose of both meat and fibre production.

Wool carpets are difficult to ignite and can achieve high levels of resistance to flammability without special treatments. The ability of a wool carpet not to show cigarette burns is considered important for hospitality end uses since the small area of char can be brushed from the carpet surface without leaving an ugly black mark. Wool is naturally soil hiding and wool carpets can contain significant quantities of soil without appearing dirty. Wool carpets respond well to cleaning and the pile often recovers well from flattening during the cleaning, thanks to the presence of the warm water used in the process. The relatively low resistance to abrasion of wool renders its use in low pile weight constructions risky, but the use of wool-rich pile fibre blends with the addition of, say, 20% nylon or other durable synthetic fibre significantly enhances the abrasion resistance of the carpet. However, the warm and luxurious connotation of wool often restricts its use to the more prestigious carpet constructions.

Polyamide

Polyamide, in the form of nylon 66 staple fibre, was introduced to the UK carpet industry in the late 1950s. Research work carried out in the laboratories of British Nylon Spinners Ltd demonstrated that a significant increase in the abrasion resistance of the wool pile carpet could be gained by the incorporation of a percentage of nylon in the pile fibre blend. The work established that the optimum blend was 70% wool, 30% nylon. Above 30%

nylon, the increase in abrasion resistance was negligible. Although different manufacturers launched products incorporating from 15% to 30% nylon in blends with wool, the most popular blend was 80% wool, 20% nylon. In the UK at this time, the industry was almost exclusively manufacturing woven carpets. The 'blending' polyamide was a premium product of 13 d'tex, slightly de-lustered by the incorporation of titanium dioxide and with a circular cross-section, and was in short supply. Since that time, however, significantly more widely available sources provide a larger range of suitable products, some of which are based on waste fibre streams and have a less defined specification.

In the USA, where the tufting process was already well advanced, DuPont and Monsanto introduced bulked continuous filament nylon yarns supported by strong brand identities. It was not until the early 1960s, however, that such yarns were introduced to the UK tufting industry. Initially, they were of 4060 d'tex with 204 filaments (based on 3 ends of 1155 d'tex, 68 filaments processed together) and had a brand requirement that the products made from the yarn should have a minimum total pile weight of 712 g/m^2 (21 oz/yd^2), but the competitive nature of the market demanded yarns of lower total d'tex for the production of carpets with lower total pile weights. These yarns had so called tri-lobal cross-section filaments which were partly de-lustered and the yarn had a low twist level of 23 turns per metre to give some cohesion to the yarn during processing. The yarns were used in loop pile carpets with a sculptured high and low loop design and were piece dyed. Following on from US developments, and to increase patterning scope in carpet, dye variant yarns were introduced such that standard, deep, extra deep and basic dyeable versions were developed. The use of cross dyeing techniques in one dyebath allowed the skilled manufacturer to obtain a wide range of colours.

Simply explained, a carpet composed of different dye variant yarns when dyed with a disperse dyestuff would be overall plain. Add to this an acid dyestuff, then the different dye levels of the yarn allows tonal and simple contrast colours to be achieved. A basic dyeable yarn would be effectively resistant to acid dyes and the acid-dyeable variant yarns would be effectively resistant to basic dyestuffs so that the further addition of a basic dye would increase the scope for high contrasts in a different colour.

Significant changes to the styles required in the market place render this technology little used by the industry at present. The most significant changes in the use of nylon yarns since their introduction to the carpet industry are as follows:

- The removal of manufacturing license restrictions in the mid 1960s, allowing nylon to become available from a wide number or sources.

- The growth in carpet manufacturers who extrude their own nylon yarn, particularly in the USA and Benelux countries.
- Developments in the production of bulk continuous filament (bcf) yarns which now enable plain cut-pile carpets to be produced from them; something that was thought near impossible 15 years ago. Consequently the call for 100% nylon spun yarns has virtually disappeared.
- The growth of nylon 6, particularly for the volume end of the market and for yarns extruded by small concerns.

Synthetic fibres offer themselves to fibre engineering to achieve additional desirable properties. In the 1960s and 1970s there was a move to make nylon more wool-like from the points of view of processability, handle and appearance. This was achieved through the introduction of a fibre that was a blend of different fibre d'tex, variable staple length and subdued lustre. This fibre had only limited success in woven carpets for which it was intended, being limited to low specification Axminster constructions. The high fibre strength of nylon posed a problem in this end use since most pile yarn for woven carpets is spun on the woollen system. Woollen spun yarns contain many fibres considerably longer than the tuft length and which may be only insecurely anchored in the tuft and which can work their way out of the tuft in use. Such fibres that remain anchored in the tuft will lie on the carpet surface giving rise to a problem called 'cobwebbing'. The use of semi-worsted spun pile yarns, which have more or less parallel fibres roughly similar in length to the length of the tuft, do not give rise to a 'cobwebbing' problem in carpet but such yarns are less bulky and less wool-like in handle and appearance.

Basic nylon fibres are similar in appearance to glass rods and have poor soil hiding performance. In fact, they are said to magnify the soil particles and appear far more soiled than they need. Several methods have been introduced over the years to give nylon increased soil hiding properties, which have varied from the adding of increased delustrant to make the fibres more or less opaque, the modification of the fibre cross-section to introduce light scattering properties, and the creation of micro-voids within the fibres. Probably the most successful soil hiding nylon is of a more or less square cross section with voids running along the fibre length.

Carpets with nylon pile were also found to generate static charges on a body walking on the carpet (more so than with traditional wool pile) which, when discharged against a conductive earthed object, caused an unpleasant shock. A method of controlling the build-up of static charge has been the inclusion of a low percentage of viscose rayon in a nylon fibre blend, on the basis that such fibres absorb moisture from the atmosphere, increasing the conductivity of the carpet surface which dissipates the induced charge. A further development has been the inclusion of a very low proportion of conductive fibre in spun nylon yarns. The problem with this was the uneven distribution of conductive fibres and the propensity for the conductive fibres to darken or dull the colour of the carpet. With continuous filament yarns, a spun yarn containing conductive fibres, or even conductive filament yarns, often filaments coated with or incorporating carbon, has been used.

Polyester

Polyester pile fibres are usually restricted to long pile styles with pronounced tuft definition, known as Saxony styles. The ability of polyester yarns to readily take a twist set makes them very suitable for such styles, which are particularly popular in the USA.

Polyester fibres have good abrasion resistance and reasonable recovery from flattening properties. Their major drawback is that they are difficult to dye and are frequently dyed under pressure. A significant advantage of increasing importance is that a valuable source of raw material comes from used, clear beverage containers, which can be ground, re-polymerised and extruded into fibre.

A particularly sophisticated process has arisen, particularly in the UK, where a small proportion, typically 5%, of low-melting point polyester fibre is incorporated into a pile fibre blend. Yarns containing such a fibre blend will have a matrix of polyester fibres which, when subjected to heat in subsequent processing, including dyeing and carpet finishing, fuse at the fibre crossing points and form what can best be described as a scaffold structure within the yarn. Such yarns tend to resist untwisting of the tufts in use, enhancing appearance retention such that singles yarns may be used. The appearance retention of carpets with pile of such singles yarns often approaches that of a more conventional pile fibre blend carpet with a two-fold pile yarn. Yarns containing low-melt polyester fibres also tend to shed less fibre in use.

Polypropylene

Polypropylene is a low cost fibre and is principally used in low specification carpets. It is characterised by its hydrophobic nature and cannot be dyed from an aqueous dye bath. Dyeable polypropylene has been developed but is not widely used because the added cost of the fibre is unacceptable for the low price carpets in which polypropylene is used. Polypropylene is coloured by the addition of pigments to the polymer before the yarn or fibre is extruded. Coloured polypropylene staple fibre is used in blends with wool to achieve heather mix effects, when the wool is dyed conventionally, in a range of shades from one yarn feedstock.

Another popular use is in bcf yarns, where different coloured polypropylene yarns are air-entangled to achieve heather mix or Berber effects. By substituting some polyamide yarn for polypropylene before air-entangling, dyeable heather mix effects may be achieved.

A significant drawback of polypropylene as a pile fibre is its poor recovery from flattening. When polypropylene is used in tight, low loop pile constructions, this is of no importance because there is little pile to flatten, but in longer pile styles, say greater than 5 mm, then the relatively rapid loss of appearance (particularly in cut pile) is acceptable only because of the low price of the product.

The increasing popularity of wood and laminate floors in recent years has created a market for rugs and carpet squares. Very attractive rugs and squares are woven on advanced face to face Wilton looms with bcf polypropylene pile.

Polyacrylate

Once seen as a wool substitute in woven carpets, the popularity of polyacrylate pile fibres has decreased significantly in recent years and is currently of only minor importance.

Vegetable fibres

Coir fibre is popularly used in so-called coconut matting.

Both cotton and jute have been used as pile fibres in conventionally constructed carpets of low importance but cotton remains a popular fibre for bath mats.

Of increasing importance in recent years however are sisal and sea grass. These are used as the face fibre in so called flat-woven products, such products being popular with many interior designers. Sea grass, however, has poor resistance to abrasion. With both face fibres, the effect of water on the appearance of the product is unacceptable as spillages of even clean water can result in staining, and wet cleaning is not possible for the same reason.

5.5.2 Fibres utilised in the backing structure

Jute

Jute is ideal as a carpet backing fibre. It is fairly inexpensive and is relatively inextensible. A particular drawback, however, is the long supply route, mainly from the Indian sub-continent, and the uncertainty of consistent supply. Jute is also subject to bacterial attack, particularly if wetted.

Jute yarn has been the most popular choice for the weft of woven carpets and many woven carpets continue to use jute yarn for this purpose. Jute is also frequently used as the 'stuffer' warp in Wilton carpets.

Woven jute fabric (hessian) is frequently the secondary backing fabric of choice because of its low cost, good dimensional stability and natural appearance. Unreliable supply has however affected its popularity for many manufacturers.

Cotton

Cotton yarns have traditionally been used as the warp yarns for woven carpets Cotton fibre, in blends with synthetics remains popular for this end use.

Polyester

Spun yarns of high tenacity polyester have been used in recent years as warp yarns in woven carpet but a more common use is as a cotton/polyester blend in chain warp yarns.

Polypropylene

Split film polypropylene yarns have been used as both warp and weft yarns in woven carpet production. As warp yarns in the form of nominal 1000 d'tex yarn as a cotton substitute for Axminster weaving polypropylene offered a less expensive alternative to cotton but led to lower weaving efficiency since the abrasion of the yarns by the reed and other weaving components caused the yarn to fibrillate and entangle.

As a weft yarn, often pigmented to a natural jute colour, polypropylene is used in weft yarns where high strength and comparatively low cost are advantageous. The yarn is less tension stable than jute but is unaffected by water and does not support bacteriological attack.

Others

Historically, because of jute shortages following the Second World War "kraft" paper was used as a substitute. This yarn was effectively twisted strips of brown paper. Its use did not continue once jute became available again.

Linen yarn has been used as the weft in some high density 3-shot wilton products necessitating a finer yarn of adequate strength.

5.6 Methods of carpet construction

5.6.1 Woven carpets

Wilton

Wilton carpets are characterised by the pile yarn being woven as warp which, when required to form the pile, is lifted above the level of the backing weave. On single face Wilton looms, the pile is usually formed over pile wires, flat metal strips which are oriented in the weft direction. The dimension of the wire determines the pile height. The wires are inserted from the right-hand side of the loom as the back pick is inserted, there being usually two or three picks per pile row, all contained in one shed of the chain warps. A number of pile wires are woven into the carpet and the one furthest away from the fell is withdrawn and reinserted as part of a new row, the wires being constantly recycled as weaving continues. The left-hand ends of the wires may be plain, so that loop-pile (Brussels) carpet is woven or have a formed end holding a replaceable sharp blade when weaving cut-pile carpet.

Conventionally, looms have a jacquard capable of selecting from five warp yarns (referred to as frames) in each pile column, but simpler looms equipped with a simple cam and lever mechanism are in use for the production of plain carpet. For plain carpets, the pile warp may be fed from a beam, but multi-frame carpets have the pile fed from a creel.

Modern Wilton looms weave two carpets simultaneously in face-to-face mode and are particularly important for the production of rugs and squares. The two carpets are woven one above the other, with the pile passing from one to the other, forming a sandwich. A knife traversing from side to side of the loom cuts the pile and separates the two carpets. Weft insertions on these looms use flexible rapiers and operate at significantly higher picking speeds than old looms using shuttle weft insertion.

Axminster

Axminster weaving processes are used for the production of patterned carpets with potentially, a greater number of colours than in Wilton carpets. There are three variants of Axminster carpet, Gripper Jacquard, Spool, and Spool-Gripper: the dominant one is now the **Gripper Jacquard** weave. The essential characteristic of Axminster weaving is that the pile yarn is inserted into the backing weave a row at a time and is the only pile yarn in the carpet, unlike patterned Wilton weaving, in which the pile yarn not selected by the jacquard to form the pile tufts lies 'dead' in the backing weave.

The pile for Gripper Jacquard weaving is held in a creel. The jacquard, commonly, has a capacity for controlling eight frames (colours), but twelve frame jacquards are also available. The ends of yarn are fed into a carrier, one carrier for each dent in the reed. Each carrier can be likened to several small, narrow boxes placed on top of each other. Each box contains a spring, and one colour of yarn is threaded into each box and is prevented from being pulled out of the box by the spring. There will be eight boxes in each carrier on a loom weaving eight frames. From the front of the loom, the carriers present eight levels of cut yarn ends.

Across the width of the loom is a set of grippers, one gripper for each carrier, often likened to birds' beaks, which swing up to the carriers to close on the protruding yarn ends. A length of yarn is drawn through the carriers, cut off, and the grippers swing down and the tufts are secured in the ground weave by a weft pick. Weft insertion is conventionally by a rigid rapier, which introduces three double stranded weft picks (shots) per pile row.

The action of the jacquard is to lift each carrier separately by the appropriate number of levels for the grippers to take tufts from the top (rest) position of each carrier. Most Gripper Jacquard looms have a pitch of 7 per inch (27.6 per dm), although variants of 31.5 and 35.4/dm are in operation. Low specification constructions may have tuft densities of 650 tufts/dm^2 whilst tuft densities as high as 1240/dm^2 may be achieved.

The basic design of the Gripper Jacquard loom has seen improvements in recent years. These include the use of electronic jacquards eliminating the need for the costly punched jacquard cards, the use of projectile weft insertion, and the use of carriers that operate in the horizontal rather than vertical orientation, allowing the grippers to operate through a much reduced arc. Loom speeds have risen considerably through these improvements.

The **Spool** Axminster process has the capacity for producing carpets with designs of an almost unlimited number of colours. The process is more labour intensive, however, and is now fairly rare. The pile yarn for Spool Axminster weaving is wound onto spools, each spool having a number of yarns wound side-by-side, like a small beam. The process of winding the yarn onto the spools is called spool setting. One spool is prepared for each row in the design repeat, with the sequence of coloured ends corresponding to the sequence of coloured tufts for that row of the design. The spools are usually either 27 inch (0.69 m) wide or 36 inch (0.91 m) wide. For broadloom weaving, a number of spools are employed across the full width of the loom. Each spool is numbered in sequence, corresponding with the number of the row of the design, and the spools are mounted in spool tube frames that allow the spools to revolve. Yarn ends are threaded through individual short tubes mounted on the spool tube frame.

The Spool Axminster loom has an endless chain in a gantry above the loom with the number of links in the chain corresponding to the number of rows in the design repeat or multiple repeats. During the weaving operation, the chain is advanced one link at a time for each row of the design, and the spool and spool frame at the weaving point are removed from the chain and lowered down into the backing weave, with each tube passing between warp ends and the pile ends gripped by a weft shot. The spools rise a short distance to draw off yarn for the row of tufts, and the yarn is cut by a pair of horizontal guillotine blades stretching the full width of the loom before the spool and spool frame are swung back into the vacated link in the chains. The ends of yarn are combed up to form a U tuft around the weft shot at the same time. On a spool Axminster loom, weft insertion is by two rigid rapiers, one above the other, which in one traverse and withdrawal of the loom insert three double stranded weft shots for each pile row of carpet.

Spool Axminster looms have a pitch of 27.6/dm and weave tuft densities from as low as 430 tufts/dm^2 up to 980 tufts/dm^2. A small plant of unique Spool Axminster looms remains in operation with a pitch of 37.4/dm, weaving carpet with a tuft density of 1470 tufts/dm^2.

The **Spool-Gripper** loom was developed to combine the patterning potential of the Spool Axminster loom with the robust weave structure of the Gripper Jacquard loom. Spools identical to those used for Spool Axminster are used and are also mounted in an endless chain. The loom is fitted with grippers identical to those used for the Gripper Jacquard loom but their arc of operation is significantly reduced. The spools remain in the chain during the weaving operation.

Textile floor coverings without pile

There has been increased interest in so-called flat woven carpets in recent years although none are produced in the UK. These carpets do not have a pile in the strict sense of the word but utilise relatively thick warp yarns which ‘float’ on the fabric surface.

5.6.2 Carpets that are not woven

Tufted carpets

Tufted carpets came about as a logical progression of the mechanisation of candlewick production developed in Dalton Ga, USA. The basic process is relatively simple but modern developments have resulted in very sophisticated machines. A tufting machine consists, in its simplest form, of yarn feed rollers extending the width of the machine, guide bars to keep individual ends of yarn from entangling, a set of needles extending across the full width of the machine and a set of hooks which interact with the needles to form the pile. The needles, which have eyes near their points, reciprocate vertically and penetrate a horizontal backing fabric tensioned between and advanced by spiked rollers. The hooks, one for each needle, are positioned beneath the backing fabric and interact with the needles when in their lowest position, to form loops around the hooks. As the needles rise, the hooks withdraw and release the loops before the cycle is repeated at high speed. The pile yarn is usually fed from large creels through plastic tubing. The tubing avoids the necessity for yarn tensioning and eliminates entangling of the yarns. Beams can also be used to feed the yarn to the tufting machine and allow shorter runs of a product to be produced more economically than from creels.

For loop pile tufting, described above, the noses of the hooks point in the direction of travel of the backing fabric, but different hooks pointing towards the direction from which the backing fabric is fed are used in the production of cut pile tufted carpet. Consequently loops are retained on these hooks and slide along them until cut by flat blades exerting a scissor action against the side of the hooks.

Tufting machines are characterised by the gauge of the needle bar and whether they make cut or loop pile. The gauge describes the distance apart of the needles mounted in the needle bar and varies typically between 3/8 inch (10.5/dm) to 5/64 inch (50.4/dm) and finer but common gauges for broadloom carpet are 5/32 inch (25.2/dm), 1/8 inch (31.5/dm) and 1/10 inch (39.4/dm). The simplest tufted carpets are limited to plain or striped patterns but sophisticated patterns and textures are now possible.

For loop pile carpets designs and textures can be produced through controlled multiple pile heights. Initially the devices to achieve this were mechanical or electro-mechanical and were limited to two or three pile heights. Such pattern attachments were also fitted to machines equipped with special hooks incorporating a spring device to produce cut and loop pile textures, the high pile being formed into cut-pile and the low pile being formed into loop-pile. The most recent advances in this field utilise servo-motor driven, multiple yarn feed rolls which allow the production of accurately controlled multiple pile height loop-pile carpet.

From the early days of tufting, the ability to move the backing cloth sideways was used to create a wave-line effect in an otherwise striped design. The ultimate development of this is the double sliding needle bar mechanism, which employs two needle bars, each of twice the machine gauge and with needles staggered so that the needles of the second bar fall in the gaps created between the needles of the first bar. Each needle bar can be accurately moved sideways between needle insertions to create designs by employing yarns of different colours threaded through the needles in a specific sequence and by controlling the sequence of movements and direction of movements of the needle bars. CAD systems are used to develop the designs and to create the instructions for the sequence of yarn colours and needle bar movements. Multiple pile height loop pile patterns may also be combined with sliding needle bar effects.

A further method of creating more sophisticated cut pile designs uses a machine with individually controlled needles which can be controlled to tuft or not to tuft for each stroke of the needle bar. Unfinished carpet already tufted on a conventional tufting machine can have a coloured design over-tufted onto it using this method. A further sophisticated development of this technology is a machine with groups of needles, each threaded with a different coloured yarn which tufts sideways before advancing the backing cloth by one row. Computer manipulation of the individually controlled needles allows the machine to produce true patterned carpet, similar in appearance to Axminster.

Patterned tufted carpet is also produced by printing.

Tufted carpets need more complex finishing than woven carpets, which have good dimensional stability from the basic weave. Tufted carpets, however, have very poor dimensional stability, are very soft and the tufts are tenuously anchored in the backing when they leave the tufting machine. A substantial latex pre-coat is necessary to anchor the tufts and to impart a stiffer handle, and frequently a secondary backing layer is laminated to achieve good dimensional stability. The laminating medium may also be latex but the latest equipment uses hot melt adhesive applied as a powder and melted before the secondary backing fabric is married to the precoated carpet. Tufted carpets finished in this way are less stiff and are claimed to be much easier to install.

Needled floor coverings

Felts have long been used as alternative floor coverings to carpets, but developments in needling thick webs of fibres has allowed a more rapid method of achieving felt-like floor coverings which are not reliant on the felting properties of wool. The simplest will have single or multiple layers of coloured fibres, with or without a carrier fabric, which are entangled by the action of closely packed barbed needles. The fabric produced is stabilised by a back-coating or a full impregnation of synthetic latex. The simplest products produced using this technology resemble flat felts, with surface interest created by different coloured fibres blended together.

More sophisticated products with pile effect can also be achieved by carefully controlling needle size, length of needle stroke and the placing of the needles in the needle board. Loop pile effects, ribs and textures are achieved in this way and a longer pile version, resembling cut-pile velvet is also possible.

Other methods

Fusion bonding of carpets has been discussed as a coming technology for very many years. However, this technology has, so far, made little impact. The process may use yarns in a warp form or a continuous web of fibre. This is adhered to a backing fabric coated with an adhesive. Typically a face-to-face process is used, with pile yarn being alternately adhered to opposite fabrics. The adhesive used may be a hot melt adhesive, activated just before meeting the pile, or a more conventional adhesive which will need heat curing. As in face-to-face Wilton weaving, the two cut-pile carpets are separated by a traversing knife.

The carpets produced by this process are often of high quality with a very even surface finish. Many variations of the fusion bonding theme have been proposed and used from time to time but the one described above is currently in production in Western Europe.

Warp knitting has also been proposed as an alternative method of manufacturing carpets. Pile fabrics have been produced, possibly more suited as rugs rather than broadloom, but there has been no significant widespread use of this technology.

5.7 Future trends

5.7.1 Fibres and yarns

There have been few significantly new fibre types for carpet manufacture since the advent of synthetic fibres many years ago. Just two spring to mind.

Poly (trimethylene terephthalate) was introduced about ten years ago and was said to combine the desirable properties of nylon and polyester fibre. In spite of this, the fibre appears to have had little impact on the market.

Reports in the technical press have referred to a new pile fibre based on a *protein polymer*. Little information is available concerning the important properties of abrasion resistance, recovery from flattening and dye-ability. Instead, the emphasis has been on the bio-degradability properties and therefore ultimate low impact upon the environment when disposed of as end of life waste.

5.7.2 Manufacture

There has been no radically new method of carpet manufacture introduced for some considerable time; instead manufacturing methods have been refined and speeded-up. The following most recent significant advances, already discussed, have been:

- Electronic jacquards for carpet weaving
- High speed gripper Axminster looms
- Continually improving face-to-face Wilton looms
- True patterned tufting, which arrived when the current demand for patterned carpet significantly declined
- Advances in multi-pile height patterning for tufting, allowing more and better defined textures and patterns.

Without a functioning crystal ball to predict the unexpected, one can only predict continued refinement of existing manufacturing processes with emphasis on speed of production, patterned carpet capability without the need for very large amounts of individual colours, and the ability to economically produce unique styles in modest quantities.

5.7.3 Influences

The importance of minimising the effects of climate change will be the most significant driving influence on all industry for the foreseeable future. Emphasis will be increasingly placed upon:

- sustainability of raw materials
- limiting the need for global transportation of raw materials and finished goods
- manufacturing with low consumption of energy and natural resources
- production of finished goods with the minimum of added chemicals and the avoidance of toxic chemicals and
- strict control of all emissions and effluents to air, land and water.

The author is also concerned about the potential damaging influences of Standardisation. Standardisation ought to be of great benefit to industry and the end user of the product. However, for the carpet industry, the end product is considered to be a construction product by Standards Organisations. This may well be relevant for carpet for commercial use but ignores the fact that very significant volumes of carpet are sold for domestic use where they are considered to be a soft furnishing. Standards rightly considered as necessary for construction products constitute an unnecessary burden on manufacturers of domestic products.

5.8 Sources of further information and advice

The following sources of information and advice are strongly recommended:

Trade associations

The Carpet Foundation (UK)	www.carpetfoundation.com
European Carpet and Rug Association	ecra@euratex.org
The Carpet and Rug Institute (USA)	www.carpet-rug.org
UFTM (F)	www.moquettes-uftm.com
VNTF (NL)	www.textielnet.nl/partners/vntf
Febeltex (B)	www.febeltex.be

Testing laboratories and technical advice

The British Carpet Technical Centre	www.bttg.co.uk/bctc
Centexbel (B)	http://www.centexbel.be
TFI (D)	http://www.tfi-online.de

Publications

International Carpet Bulletin	www.world-textile.net

Text books

Carpet Manufacture by G H Crawshaw, Woodhead Publishing, UK	ISBN 0-908974-25-6

Part II

Developments in interior textiles

6

Advances in joining fabrics for the furniture industry

E. STRAZDIENE,
Kaunas University of Technology, Lithuania

Abstract: An overview of stitch types, an analysis of industrial needle construction and a classification of sewing seams is presented. Technological advances are described on the basis of the latest developments of sewing and sealing machines. Special attention is paid to the quality of formed seams from the standpoint of seam puckering. Peculiarities of leather joining are analyzed. Also, examples of hot-air sealing machines for continuous seam-sealing on water resistant, waterproof and breathable materials are presented. Innovative laser technologies applied for such products as auto and furniture upholstery are briefly discussed.

Key words: sewing machine, stitch, seam, needle, puckering, seam-sealing.

6.1 Introduction

Upholstered furniture relates to us directly because, throughout history, it has expressed the status, the outlook and the ambitions of its users most obviously. Upholstery, in its traditional meaning, describes the use of textiles in the complete furnishing, i.e. in bedding, seating, floor coverings, drapery, etc. The Encyclopaedia of Furniture Materials, Trades and Techniques[1] gives 'furnitures' as a general term to include all textile fabrics used in bed and upholstery work, i.e. brocades, cretonnes, reps, tapestries and many others. Meantime James[2] has noticed that today the word 'upholstery' has become synonymous with chairs and seating, while the term 'soft furnishings' deals with all other fixed fabric furnishings, especially drapery and detachable coverings for windows and beds.

Joining of covers and fabrics is a specific feature of modern upholstered furniture. Industrial sewing is the most common manufacturing operation for this purpose because textiles are a major integral part of the upholstery process. Modern sewing machines, basically lockstitch, are designed to deal with such specific tasks. This chapter is focused not only on joining operations of traditional textiles, but also on heavy industrial sewing, such as sewing of automobile seat cushions, etc. The latter requires not only high productivity, but also high sewing quality,[3] i.e. good appearance and long lasting stitches. Typically, the material being sewn in this case includes single and multiple plies of synthetic fabric, leather or coated fabrics which are intended to be an imitation of leather.

Particular emphasis is laid on leather joining, which needs special threads, needles and joining equipment. Leather is a generic term for large animal hides and small animal skins from reptiles and birds, prepared by a currier. It has been used by furniture makers for many functional and decorative tasks, which include seat and upholstery covers, cushions, wall hangings and writing surfaces. Thus, leather remains an important upholstery and decorative finish for all sorts of furniture.

For centuries, needlework based on passing a needle and thread through a cloth was the main technique for joining operations. Though hand stitching became masterly, it was in the 19th century that the sewing of covers was aided by the use of the first industrial sewing machine. The development of the machinery since then has seen continuous improvement in speed, quality and complexity. The use of semi-automatic, automatic, numerically-controlled and computer-programmable machines is now standard in the furniture industry. Contemporary upholstery work has ranged from the practical to the highly experimental. The exploitation of materials such as stretch fabrics for the creation of sculptural shapes and the imaginative use of foams for fantasy furniture are interesting developments for the future.[1]

6.2 Principles and types of joining methods

Nowadays, the principal method of joining upholstery materials and fabrics is machine sewing. During the last forty years, the design and application of sewing equipment has changed and developed from standard drop-feed lockstitch machines to very specialized and sophisticated machinery. The most innovative areas of upholstery fabrication involve advanced sealing or welding equipment based on high frequency, laser etc. technologies.

Stitch formation is the basis of the sewing process. According to standard ISO 4915, the term stitch is defined as 'one unit of conformation resulting from one or more strands or loops of thread intralooping, interlooping or passing into or through material. A stitch can be formed without material, inside material, through material or on material'.[4] A distinction between the terms intralooping and interloping is made to clarify the way in which thread is arranged in the stitch. Intralooping is the passing of a loop of thread through another loop formed by the same thread, whereas interlooping is the passing of a loop of thread through another loop formed by a different thread. Interlacing refers to passing a thread over or around another thread or a loop formed from another thread.[4]

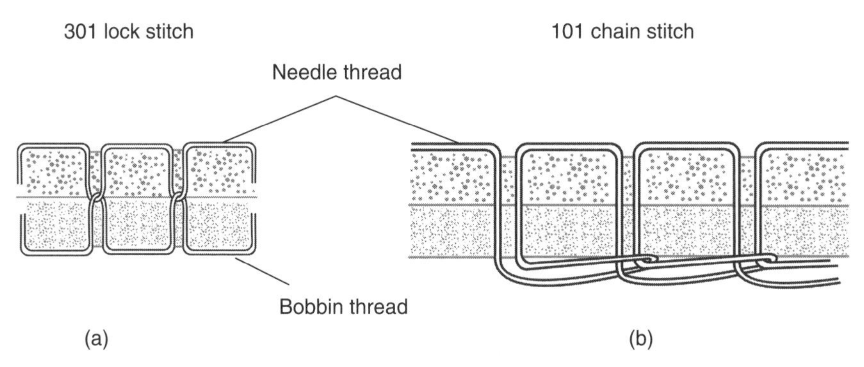

6.1 Lock stitch (a) and chain stitch (b) formation principles.

6.2.1 Stitch and seam types

Machine lockstitch has been used for sewing fabric plies together for as long as the sewing machine has existed. A modern lockstitch sewing machine has added convenience, strength, and flexibility, but it still works by the same principle. Lockstitch is the strongest seam stitch that can be produced.[5] The lockstitch machine uses a needle thread and a bobbin thread, which interlock in the centre of the joint (Fig. 6.1a). In order to obtain qualified seam, fine adjustments are necessary, during which the tension on both threads is set so that locking takes place inside the plies and not on the surface of either. At the same time, the stitch length must be adjusted to suit the particular fabric or material. The size of machine needles and the thread being used must be compatible, and selected to suit the materials being sewn.[6] Needle thread is exposed to varying stresses, strains, heat, and abrasion, while the bobbin thread is relatively inert.

Machine chain-stitch formation essentially differs from a lockstitch. It is not interlocking but interlooping. The simplest stitch is formed by a single-needle thread (Fig. 6.1b). More complex and stronger seams are produced with two and three thread stitch formations, all of which are applied by needles and loopers. Although the stitch produced is not as strong or as durable as a lockstitch seam, it does have some advantages, for example high elasticity[7] and is used increasingly in furnishing and upholstery for preparation work, e.g. in the making of pipings, zip insertion and other pre-sewing operations. Fast continuous sewing makes chain-stitch seaming attractive for modern upholstery manufacture.[2]

According to standard ISO 4915, stitches are classified into six classes, which are identified by the first digit of a three-digital number. Each particular stitch type is designated and identified by the second and third digits.[4] Classes are distinguished by the geometrical arrangements of threads in the stitches and a number of stitch types:

- Class 100 features stitches most oftenly formed with one intralooping needle thread;
- Class 200 stitches can be formed by hand, although nowadays in most cases they are formed by machine;
- Class 300 stitches are formed with two or more threads or groups of threads and are characterized by interlacing of two or more groups;
- Class 400 features multi-thread chain stitches which are formed with two or more groups of threads and are characterized by interlacing and interlooping of the two groups;
- Class 500 covers overedge chain stitches, which are formed with one or more groups of threads and are characterized by loops from at least one group passing around the edge of the material;
- Class 600 includes covering chain stitches, which are formed with at least two or three groups of threads and are characterized by two of the groups covering the raw edges of both surfaces of the material.[8]

No matter how sophisticated or well-developed an industrial sewing machine appears to be, the basic forming of a stitch remains the same. The basic stitches used in the furniture industry are: lockstitch, single thread chain-stitch, two thread chain-stitch, overedge chain-stitch, and overedge with safety stitch. The most widely used stitch type, and certainly the best for upholstery, is the lockstitch (Class 301). It is universal and almost the strongest seam stitch that can be produced. Table 6.1 presents not only the most recommended types of stitches for textile joining in the furniture industry, but also the preferred stitch densities for each stitch type and average thread consumption by needle (NF) and bobbon/looper (GF).[6] At the areas of furniture where higher strength and elasticity is needed, safety stitch is applied, i.e. an additional chain-stitch (Class 401) is made a few millimetres inside the overedge seam (Class 503). In such cases, two seams are produced at the same time, but they are independent.

Standard ISO 4916[9] defines the seam as 'the application of a series of stitches or stitch types to one or several thicknesses of material'. In the other sources, seaming has a wider description and is sometimes defined as the joining of one or more plies of material, achieved by stitching, welding, adhesives or by other means. According to the latter description, a seam consists of the agent which holds materials together (e.g. thread, resin) and the materials being joined (e.g. textile, film) which contains that agent.

Referring to ISO 4916, each stitched seam is designated numerically by five digits:

0 – shows the class from 1 to 8;
0.00 – shows the material configuration from 01 to 99;
0.00.00 – shows needle penetrations and/or material configurations from 01 to 99.

Table 6.1 Amann sewing thread requirement table

Stitch class	Stitch name	Seam width (mm)	Stitches (per cm)	Thread required	
				(per 1 m of seam)	%
ISO 101	Single thread chain stitch	–	2	NF – 3.80 m Total: 3.8 m	100
ISO 301	Two thread lockstitch	–	4	NF – 1.4 m GF – 1.4 m Total: 2.80 m	50 50
ISO 401	Two thread chain stitch	–	4	NF – 1.7 m GF – 3.1 m Total: 4.80 m	35 65
ISO 304	Two thread zig-zag lockstitch	5	4	NF – 2.7 m GF – 2.7 m Total: 5.4 m	50 50
ISO 504	Three thread overedge stitch	5	4	NF – 1.7 m GF – 12.1 m Total: 13.8 m	12 88
ISO 801	Four thread safety stitch (class 401 + class 503)	–	4	NF – 8.2 m GF – 8.2 m Total: 16.4 m	50 50

Thus, the seams are divided into eight classes according to the types and the minimum number of components in the seam. The components of the seam are termed as being of 'limited' or 'unlimited' width, indicating possible constraints imposed on selection of stitch and/or seam types and the arrangement of material plies at joining.

According to James,[2] the main types of upholstery seams are:

- plain seam (Class 1.01.01), which is the main jointing method for all covers and produces a strong face-to-face joint (Fig. 6.2a);
- reinforced seam (Class 2.02.03), which is plain-sewn and topstitched; it produces a very strong seam that looks best on plain-woven flat cloths, coated fabrics and hides (Fig. 6.2b);
- piped or welted seams (Class 1.15.02), which are plain seams with corded or uncorded piping inserted between the plies; they produce a bold and strong joint on most upholstery loose covers and boxed cushions (Fig. 6.2c);
- double-fell seam (Class 4.03.03), usually topstitched on a twin-needle machine; produces a decorative joint applied on most covers to flatten

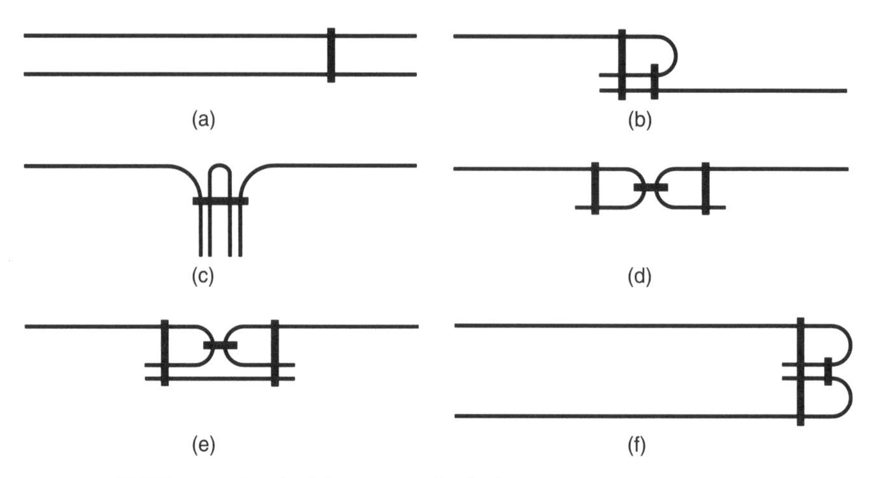

6.2 Types of upholstery manufacturing seams.

plain seams; it looks best on coated fabrics, hides, and flat, plain-woven covers (Fig. 6.2d);

- reinforced taped felt seam (Class 4.07.04), which is a twin-needled plain seam with a reinforcing tape; it is a strong and decorative seam applied mainly where strength is needed on flat areas and around curved foam-covered edges (Fig. 6.2e);
- top-stitched edge seam (Class 1.06.02), a plain-sewn decorative edge seam which is over-stitched after reversing; it produces a heavy bold edge and supports edges of soft covers and soft hides. (Fig. 6.2f).

Modifications of above mentioned are gathered seams which are applied to form gathering of different densities as a decorative effect or as a functional seam along edges and around cushions. Overedge seams are used to oversew the edges of fabric plies and to produce a safety seam parallel to the overlocking. This seam is recommended for knitted or stretch fabrics because it is very flexible. Mainly, it is applied on loose covers, bed covers and washable cushion covers. In some specific furniture manufacturing operations, hand stitching is needed, e.g. lock stitch, running stitch, over-sewing or slip stitches.

6.2.2 The strength of stitched seams

Usually a sewing seam is selected on account of its quality and decorative appeal. In addition, attention must be paid to its strength. There are five main factors that determine the strength of a sewn seam: fabric type, thread type, seam and stitch type, thread tension, and sewing machine needle type.

The first factor – fabric type – affects seam performance, depending on fabric content. Fabric construction, e.g. woven or knit, type of weave, weft count, yarn type and size, as well as seam direction and propensity of the yarns in the seam to shift or pull out, also plays a significant role for seam strength.[10,11]

The second factor – thread fibre type, construction and size – has a definite effect on seam properties, e.g. a spun polyester thread gives greater seam strength than a cotton thread of the same size. Sundaresan *et al.*[12] reported investigations which indicated that cotton threads exhibit greater strength loss owing to their poor abrasion resistance; and that comparatively shorter fibres in cotton threads are responsible for the higher strength reduction. With regard to thread construction, core threads made with continuous filament polyester core generally provides higher seam strength than spun and textured threads. This also relates to thread finishing because, in most cases, it determines thread friction coefficient and the thread's abrasion resistance.[13]

The third factor – stitch and seam construction – is very important. Seeking for better seam selection, Masteikaite[14] investigated fused and non-fused seam properties in respect of the indicated indexes, i.e. the number of fabric layers in the seam and applied stitch and seam types. It is well known that the more thread that is consumed in a stitch, the greater the seam strength. This can be illustrated by comparing 301 lockstitch seam with 401 chain-stitch seam. Threads used in 301 lockstitch seams are more susceptible to shearing each other than 401 chain-stitch and 504 overedge seams because of the way the threads are interlocked together rather than interlooped together.

The effect of the fourth factor – stitch density – indicates that the greater the stitch density, the greater the seam strength. This phenomenon was tested by Ujevic and Kovacevic[15] on seam strength of car seat coverings. Testing samples included technical fabrics, nonwoven materials, knitted fabrics and artificial leather. The authors concluded that seam samples having shorter stitches, i.e. higher stitch density, were stronger. In their research, the densities of 5, 4 and 3 stitches/cm were investigated. This refers back to the point that the more thread is put in the seam, the stronger is the seam. However, on some fabrics, e.g. leather, too many stitches can cause damage to it by cutting. Thus, for decorative leather seams, larger stitch lengths can be used and are very effective; 4 stitches per cm is the normal minimum in this case.[2] Excessive stitch density can also contribute to seam puckering and reduce the quality of final products.

The fifth factor – stitch balance – can be accomplished by adjusting the sewing machine thread tensions, thread control guides, and eyelets, etc. Care should be taken not to put too much needle thread in the seam to cause the seam to open up when stress is applied to it. Excessive sewing machine thread tension will cause reduced seam strength as well as create other sewing problems.[2]

Thus, significant work has been done related to the investigations of sewn seam strength, starting with traditional textile materials and ending with more complicated structures such as laminated textiles and composites.[16,17] Mouritz *et al.*[18] have presented a review of over fifty studies concerned with the effect of through-the-thickness stitching on the in-plane mechanical (tensile, compressive, flexure, interlaminar shear, creep, fracture and fatigue) properties of fibre-reinforced polymer composites. They noticed many contradictions: some studies reveal that stitching does not affect or can improve slightly the in-plane properties while others find that the properties are degraded. They noted that in reviewing these studies, they demonstrated that materials joining processes are very complex and that predicting the influence of stitching on the in-plane properties is difficult because it is governed by a variety of factors, including the type of composite (type of fibre, resin, lay-up configuration), the stitching conditions (type of thread, stitch pattern, stitch density, stitch tension, thread diameter) and the loading conditions.

6.2.3 Sewing machine needles

The sewing machine needle must penetrate the material without damaging it and must push the yarns aside. Also, it must protect the thread and guide it through the material, finally forming the loop. Thus, different types of sewing machine needles are available according to their application, the characteristics of the sewn material, the size of the sewing thread, the type of seam and the stitch type.

The most important constructional characteristics of a sewing machine needle are: the blade, the long groove, the scarf and the eye (Fig. 6.3). The blade of the needle runs from the end of the shoulder to the beginning of the eye. Often, the blade increases in thickness from the eye to the shoulder. This is considered as needle reinforcement, providing higher stiffness. On the threading side of the needle is the long groove. Its function is to guide the thread while forming the stitch, and to protect it against excessive friction. Above the eye there is a scarf, which facilitates the passage of the hook into the loop and reduces the danger of missed stitches. The shape of the eye is always extended in its length, because the needle thread has to pass diagonally through the needle in the length direction.[19]

The metric size (Nm) of a needle defines the diameter of the blade (in 1/100 mm) at a point just above the scarf (Fig. 6.3). Fine needles have a size up to about Nm 70; medium needles are about Nm 80 or Nm 90; thick needles have a size greater than about Nm 110. The range of needle sizes for use in upholstery sewing is generally from 100 up to 150.[2]

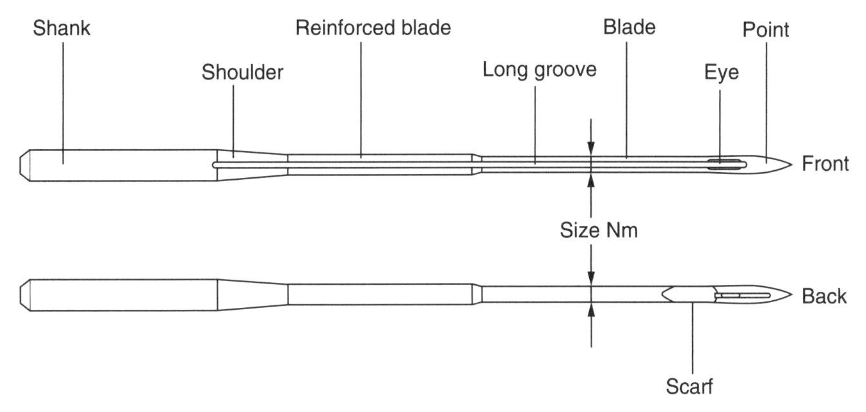

6.3 An industrial sewing machine needle.

Lomov[20] studied the interaction of a woven fabric with a needle, taking into account fabric structural parameters and warp and weft geometrical and mechanical properties. He proposed an algorithm for the computation of maximal needle penetration force. The results obtained provided the dependence of needle penetration force for a plain-woven fabric not only upon the properties of sewn materials and threads, but also upon such needle characteristics as diameter and surface angle. He concluded that the shape of the needle is of crucial importance for the quality of sewing operations.

Needles are manufactured with a wide variety of needle points appropriate for the differing properties of materials which have to be sewn. Selection of the right needle for a certain sewing operation determines the quality of the seam, as optimal design of the needle point is of paramount importance in decreasing the needle penetration force when the needle is pushed through the material.[21] There are two basic classes of points, namely Round Points and Cutting Points. Round points have a circular cross-section but may have two basic shapes known as Set Points and Ball Points, which are suited for different materials (Fig. 6.4).

Slim set point needles are very sharp and can penetrate and damage the yarns of the material. Thus, they are not suitable for knitted stretchable fabrics and are mainly used for fine wovens. The most versatile point shape – *set cloth point* – is slightly rounded and displaces the yarns of the material without damaging them. The *heavy set point* is strongly blunted. *Light ball* points are used for sensitive fabrics such as knits, to prevent damage to the loops. Elastic materials containing rubber or elastomeric threads are sewn with *medium* or *heavy* ball points. For upholstery sewing, needles with cutting points which are used for sewing leather and films or coated and laminated textiles are advised. They are classified and named according to

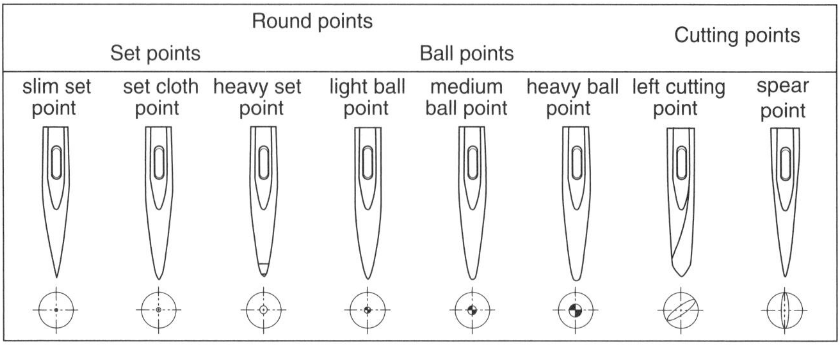

Round points
Set points
Ball points
Cutting points
slim set point
set cloth point
heavy set point
light ball point
medium ball point
heavy ball point
left cutting point
spear point

6.4 Shapes of sewing machine needle points.

the position of the cutting edge and its shape. The shapes are named with regard to the form of the cutting edge, e.g. spear point, triangular point and diamond point.[19]

It must be noted that solid materials used in the furniture industry, e.g. leather or laminated textiles, are holed during sewing. Thus, the shape of needle point is very important for efficient sewing of hides. Though leather can be well sewn with a standard round cloth-point needle, those needles specialized for leather work have different point designs. These can be spear and chisel shapes, which have a cutting action as they pass through the leather, and they are often referred to as leather points. Needle selection is also dependent on the complexity of the seams as the cutting action made by cutting point needles tends to produce stitches which have a slanting or zig-zag appearance and are better suited to decorative work than to standard plain sewn joints.

This phenomenon is also shown by an investigation during which the needle and the stitch parameters upon polymeric leather and PU film behaviour under biaxial loading were analyzed.[22] The effect of stitch density for the distribution of principal stresses was evaluated by a finite element method. The experimental part was performed with a special punching unit which simulated real conditions of biaxial loading. Such loading is much closer to real use conditions than is uniaxial tension. The radius of tested specimens was 56.4 mm and the radius of the applied punch was 30 mm. The mechanical properties of the coated and laminated leather used for these investigations were as follows:

- laminated *Permair* leather (hybrid leather obtained by hot plate pressing of a separately moulded microporous polymer film): thickness (1.71 ± 0.25) mm; tensile strength (21.9 ± 1.7) MPa; elongation at break (37 ± 3)%; Young's modulus 68.8 MPa;

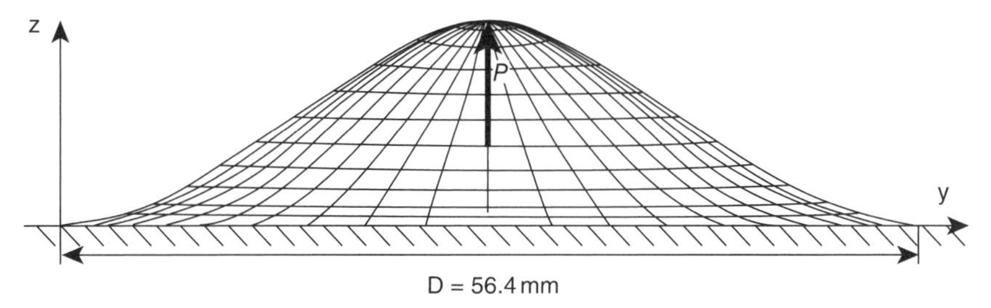

6.5 Generated meshes of shell elements used for finite element simulation of biaxial loading.

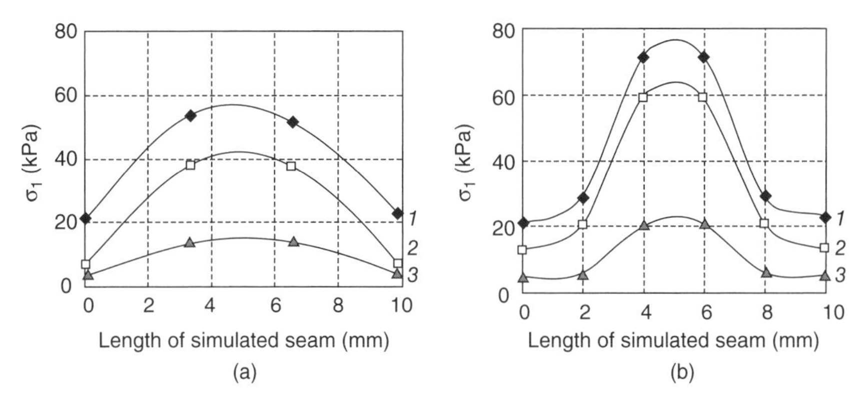

6.6 The distribution of maximal σ1, along the length of simulated seam during biaxial deformation for different stitch densities: (a) 3 cm^{-1}; (b) 5 cm^{-1}; 1 – laminated leather; 2 – pigment coated leather: 3 – PU film.

- elastic *Nappa* leather (chrome-tanned cowhide leather with pigmented coating): thickness (1.25 ± 0.05) mm; tensile strength (11.9 ± 2.1) MPa; elongation at break (41 ± 4)%; Young's modulus 36.6 MPa;
- microporous polyurethane film *Permair* (consisting of interconnected pores with a diameter of 5 μm): thickness (0.40 ± 0.05) mm; tensile strength (7.5 ± 0.2) MPa; elongation at break (326 ± 14)%; Young's modulus 18.2 MPa.

Finite element analysis was carried out, applying multi-layer four-node shell elements. The calculation algorithm was based on structural geometrical non-linearity, assuming that all components are linear-elastic. The generated mesh used for the analysis is presented in Fig. 6.5. Two different stitch densities, 3 and 5 stitches/cm, were simulated during this analysis while the length of simulated seam was 10 mm.

Stress distribution in the leathers with different finishes is presented in Fig. 6.6. Finite element analysis proved that the nature of the material has strong influence upon the stress distribution. Stresses reached their maximal value in the middle of the simulated seam. In this area, stress values are higher by about four to five times compared with those at the beginning or at the end of simulated seam. The highest stress was developed in the laminated leather. Comparison of stitch densities proved that the increase of stitch density increases maximal stresses during biaxial loading.

Distribution of principal stresses in leathers under biaxial loading depends not only upon the stitch density, but also upon the stitch hole shape. Later experimental investigations on the behaviour of the laminated leather with respect to the shape and distribution of stitches under punch loading were performed (Fig. 6.7). Results obtained revealed that punching strength slightly depends upon hole shape when the stitch density is low, especially for oval holes. The rhombus shaped holes at higher densities decrease the punching strength of leather due to the formation of defect zones with high stress concentrations. Summarizing the results, it can be concluded that increase of stitch density decreases the strength of the tested leathers in biaxial deformation, i.e. crack formation and propagation depends on the stitch density and needle shape.[22]

6.2.4 Sewing machine needle heating

In heavy industrial sewing, needle heating has become a serious problem that limits increases in sewing speed, and hence productivity. The rise in temperature of the sewing machine needle can be caused by friction with the fabric during penetration and emerging and/or by friction of the thread passing through the eye. If the friction of the fabric is low, the heating of the needle is predominately caused by the thread passing through the eye. Depending on sewing conditions, the maximum needle temperatures range from 100° up to 300 °C. According to the Rhein-Nadel company, the limit is a needle temperature of approximately 130 °C. When this limit is exceeded, the thread distributes the heat. Thus, needle heating depends on: the fabric (kind of fibre, weave, knitted goods, finish, number of plies, fabric combinations), sewing machine (speed, stitch type), thread (kind of fibre, finish, construction), needle (size, surface, point, shape) and environment conditions (humidity).[23] According to the Rhein-Nadel investigations, the highest needle temperature is usually measured at the needle eye and strong dependency exists between the distribution of heat along the needle and the penetration force (Fig. 6.8).

Many investigations have been conducted on needle heating, such as measuring the needle temperature, studying the heating mechanisms, and correlating the sewing parameters and fabric properties with the maximal needle temperature. Khan *et al.*[24] analyzed the nature of the forces acting

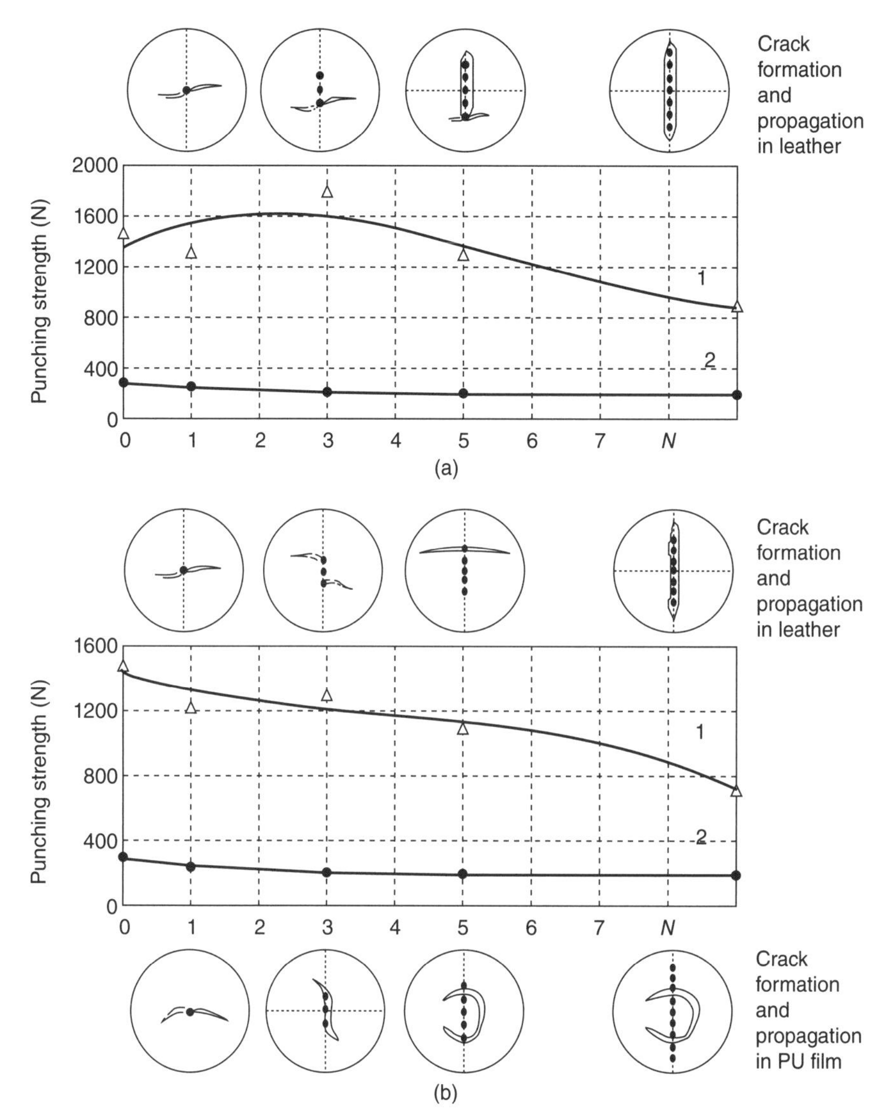

6.7 Dependence of the punching strength and crack formation and propagation upon the shape and number of needle pricks: (a) oval prick; (b) rhombus prick; 1 – pigment coated leather: 2 – PU film (N = number of holes).

in the sewing needle penetration process and identified the factors that influence the needle–fabric interaction leading to heat generation. Four major variables were considered: needle velocity, needle diameter, needle surface finish and the number of fabric layers. It was found that needle diameter affects the maximum penetration force and energy of

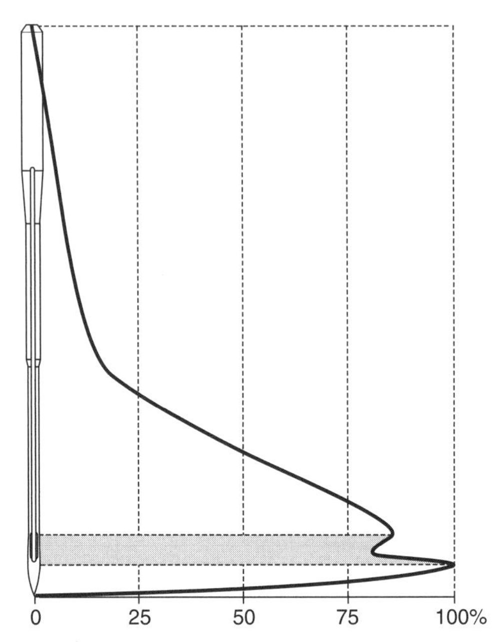

6.8 The distribution of heat along the needle during sewing.

penetration. Testing results have revealed that special high-emissivity needle-surface finishes reduce the temperature of needles during sewing. Experimental testing did not show any differences in needle behaviour for a variety of needle-point shapes, e.g. twist point, diamond point and standard ball point.

According to the literature, most needle heating studies used experimental methods based on the use of a thermal couple, temperature-sensitive materials, IR radiometry and infrared pyrometer. However, they have major limitations because they are expensive and time-consuming. In addition, they cannot directly reveal the causes of the needle heating. Experimental study of needle heating in sewing heavy materials, such as upholstery fabrics, were performed by Liasi *et al.*[25] in order to study the effect of such factors as: the shape of needle point, sewing speed, stitch density, fabrics, and thread tension. Analysis was carried out on the basis of thermal images of the needle during sewing obtained by infrared (IR) radiometry. Normal IR radiometry was used in lower speed sewing, i.e. approximately 500 rpm. High-speed IR and high-speed line scanning IR radiometry were used for medium speed sewing (1000–2000 rpm). It was found that, even with air vortex cooling, the needle can still reach high enough temperatures to affect the sewing quality and even cause thread breakage; also, that high temperatures weaken the thread, since thread tensile strength is a function of temperature.

Li *et al.*[26,27] has presented two relatively simple analytical models: the sliding contact model and the lumped variable model. Both models can predict the needle temperature rise starting from initial heating to steady state with respect to needle geometry, sewing conditions and fabric characteristics. Together with analytical models they presented a finite element analysis (FEA) mode.

A problem that accompanies high-speed sewing is broken thread, caused for various reasons. However, excessive needle heating is definitely one of the leading factors. In addition to other problems, the hot needle can cause the formation of creases in the sewing thread, making the loop formed by the sewing thread of improper shape and the hook not to enter the loop. This results in skipped stitches. The hot needle can also initiate damage in the fabric, such as leaving burn marks on natural fibres, leaving a weakened seam or melted residue on the fabric surface. Decreasing the sewing speed can reduce the needle heating. Thus, it will also decrease the productivity.

The mechanism of strength reduction of sewing threads has been discussed by Sundaresan *et al.*[28] They studied the effect of fabric tightness and certain thread properties such as its size, coefficient of yarn–metal friction, twist direction, number of piles, type of fibre and fibre denier upon strength reduction of the thread. Ziliene and Baltrusaitis,[29] in their research, once more confirmed that mechanical properties of sewing thread play an important role in high-speed sewing and that strength reduction of spun twisted and air-jet textured sewing thread is dependent upon sewing speed, needle size and fabric characteristics.

Possible solutions to overheating are: thread lubrication, dabbing or spraying the fabric with a lubricant; blowing air or air/silicone mixture against the needle, and reduction in sewing speed. These are also important objects of investigations. Michael *et al.*[30] analyzed the effect of various sewing conditions and needle characteristics on the temperature observed during high-speed sewing. The maximum needle temperatures were determined by infrared flux measurements. The temperature was found to increase with speed. Also, it was found possible to reduce the needle temperature by 10–15% using a cyclic type of operation, with short periods of sewing followed by short periods of idle time. Thread or cloth lubricants did not show sufficient effect for needle cooling. However, the use of forced air cooling was found to be very effective. The application of all the investigated factors showed that it was possible to reduce the temperature of a chrome-plated needle sewing at 4000 rpm from 700 °F (371 °C) to below 300 °F (149 °C).

6.3 Technological advances

6.3.1 Universal lockstitch sewing machine

A sewing machine consists of a casting with a head, an arm, a column and a bed. The casting incorporates the thread take-up mechanism, the hook or looper, the needle drive and the feeding mechanism. Lockstitch machines are produced as single-needle and twin-needle models. They are available in three basic forms: flat bed, post bed and cylinder bed (Fig. 6.9). A flat bed is designed for normal sewing operations, while the other two are more specialized. The head of post bed machines is elevated above the table and tilted towards the operator for easier working and vision. Usually these are twin-needle machines. The cylinder bed arrangement producers a single-needle stitch at the end of the table, with an extension table to support the work during sewing.[2]

The operation of the universal lockstitch sewing machine is based on two separate needle and bobbin threads. Both threads are fed under tension by the tension mechanism to ensure a balanced stitch. In the stitch formation process, the threaded needle goes down into the fabric, and during its withdrawal a loop of thread is formed on the opposite side of the long needle groove. A rotating hook is timed to revolve and to pick up the looped thread, which is then passed completely around the bobbin. This loop is formed due to the thread hitting the end of the eye during the upwards movement of the needle and due to the friction of the thread in the long groove. As the needle emerges from the fabric, it pulls the stitch into the place and the take-up lever tightens the needle thread.

Thus, it is evident that needle thread is exposed to varying stresses, strains, heat and abrasion. The same effects upon the bobbin thread are not so significant. Knowledge of the interactions between a sewing machine's mechanisms and the sewing thread in the stitch formation process is of great importance for the quality of sewn upholstery products. Significant amounts of research work related to the problems of thread loading and the operation of different sewing machine parts has been done over the years.

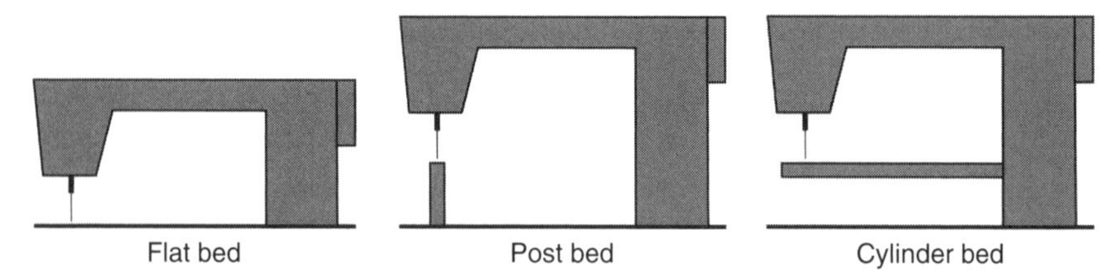

6.9 The forms of lockstitch sewing machines.

In recent times, in order to increase sewing productivity, high-speed sewing has been used extensively. Currently sewing speeds range from 2000–6000 rpm. In heavy industrial sewing, e.g. upholstery materials, typical sewing speeds range from 1000–3000 rpm.[26] Thus, the effect of a sewing speed upon thread strength is of great importance. Rudolf *et al.*[31] have studied the characteristics of 100% polyester core-spun sewing threads at different stitching speeds and have determined thread tensile forces and changes to thread thermo-mechanical properties during and after the sewing process. They proposed an analytical method which was aimed to monitor real-time thermal and mechanical effects upon structural changes in thread-twisted fibres during fabric joining.

In order to measure the variation of thread tension forces in a sewing process, the needle bar's and thread take-up lever's kinematics were analyzed by computer simulation.[32] A cyclogram was drawn which enabled analysis of the interactions between the thread and the elements of the sewing machine. This made it possible to establish the exact positions of the mechanisms with respect to the main shaft rotation in the sewing machine. Kamata *et al.*[33] paid attention to the disengagement of the needle thread from the rotating hook in an industrial single-needle lockstitch sewing machine with respect to the tightening tension. Experimental investigation revealed that the timing of the disengagement of the needle thread from the rotating hook is affected by the elongation properties of the sewing threads. Ferreira *et al.*[34] made a study of thread tensions on a lockstitch sewing machine, measuring simultaneously both the needle and the bobbin thread tensions. Experiments were carried out under specified sewing conditions, i.e. sewing speed, number of plies, fabric and sewing thread qualities, in order to investigate the effect of these factors on the needle and bobbin thread tensions. Four significant peak tensions on the needle thread tension trace and two significant peak tensions on the bobbin thread tension trace were detected during the stitch formation cycle.

Because the proper operation of all sewing machine parts is of the highest importance for the quality of produced products from the standpoint of stitch shape, there has been much research done in order to develop apparatus and devices suitable for machine mechanism performance analysis. Osawa and Saito[35] developed an opto-mechanical lever apparatus which produces feed dog motion profiles. Another projector is used to display dynamic torque profiles. These apparatuses are easily attachable to and removable from industrial lockstitch sewing machines for use as service tools.

Not only experimental studies, but also mathematical modelling of sewing machine mechanism operations is presented in literature. Zajaczkowski[36] described the behaviour of a system composed of the rotary hook and feed mechanism driven by two triangular cams, using a set of non-linear autonomous ordinary differential equations. It was found that, due to the timing belt extensibility, the actual relative positions of the needle and the hook were different from those expected. The vibration of the belt increased with speed.

Besides the traditional approach to the operating characteristics of certain sewing machine parts, more complicated research in the fields of 'intelligent sewing machines' has been carried out. The work of Stylios and Sotomi[37] illustrates this; they devised a neuro-fuzzy control model which incorporated discrimination of material characteristics to be stitched by automatic determination of their properties. The fabric/machine interactions at different speeds were computed and implemented in a neural network. The model was successfully applied to an instrumented industrial sewing machine.

6.3.2 Feeding mechanisms of sewing machines

A basic lockstitch machine can have one of several different feeding mechanisms. It should be noted that this part of the sewing machine is critical for good quality stitches and deserves attention above all other sewing machine mechanisms (Fig. 6.10).

The most common is the drop feed system (Fig. 6.10a), in which the fabric is fed forward by the moving feed dog with the presser foot held onto the fabric by spring pressure. Needle movement is in a straight line up and down, with feed taking place when the needle is in the up position. Although this system is the most conventional, it is not recommended for general upholstery sewing which is mainly in the heavy class.[2] Also, it is not recommended for slippery materials because their joining is always problematic and often results in different distortions, e.g. transportation puckering. Hence, attempts have been made to create a mathematical model of a feed mechanism which uses a feed dog covered with a supple material.[38] In the calculations, the rheological properties of the fabric were considered. It

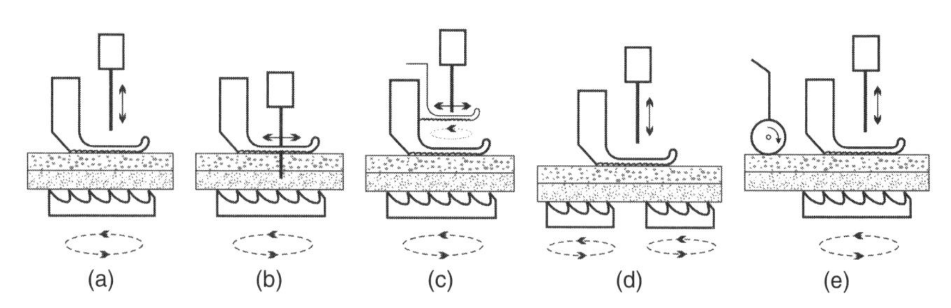

6.10 Feeding mechanisms of lockstitch sewing machines.

was found that an increased coefficient of friction and suppleness of the feed dog covering enabled proper transportation of slippery materials. Also, supple material enabled avoidance of the tooth-shaped coupling obtained by pressing the teeth into the fabric. Compound needle feed mechanism, to a certain extent, eliminates the problem of fabric plies slipping as described earlier (Fig. 6.10b). Feed in this case takes place when the needle is making a forward movement in unison with the feed dog. Thus, puckering of the bottom plies is less frequent. Compound feed with a plain presser foot is recommended for light and medium upholstery covers and some leathers. The combination of compound feed fitted with an alternating presser foot (Fig. 6.10c) is very well suited to upholstery sewing. The presser foot in this case is a two-part mechanism arranged to alternate in the pressing down of the fabric plies. This feed system, called a walking foot, is considered adequate for general and heavy upholstery work and is suited to hide work and the joining of heavy tweeds and pile fabrics.

Puller feed (Fig. 6.10e) uses a powered roller to assist feeding and can be fitted to most sewing machines. It is used mainly for long runs of border making, zip insertion and where different types of material are being quilted or laminated together. In general puller-feed systems are designed to assist in the straight sewing of difficult materials and are standard on automatic seamers[2].

One of the features of furniture manufacturing is the creation of 3D objects from 2D fabrics. Spatial shape of soft furnishing, e.g. mattresses, is obtained by applying appropriate sewing operations: puckering and gathering of fabrics. In such cases, different lengths and contours have to be fitted and often fabrics with different material properties must be joined. Sewing machines with differential feed (Fig. 6.10d) are designed principally for situations where the puckering and gathering of fabrics is required. The feed mechanism is basically drop feed, employing two feed dogs working in tandem. These two can be adjusted to work at different rates, thus forming a gathering function in front of the needle. For the gathering of heavier upholstery fabrics, a swing-out pressure plate is used. This has been found to be more effective and sits on the fabric in front of the presser foot.

Studies have been performed to clarify the effects of sewing conditions and fabric properties by easing the variable top feed of the sewing machine.[39] The results obtained were as follows: seam shrinkage tended to increase with the ratio of differential feed and important factors which influenced easing were tensile strains, bending rigidity and fabric thickness. The first two were obtained by the KES-F system along the sewing direction. Flexible automation of sewing processes with the use of a roller feeding device for sewing gently curved pieces is considered to be a more modern approach.[40]

6.3.3 Seam puckering and its factors

In general, puckering refers to the wrinkled appearance of a seam, which can be caused by a number of factors including: feed puckering, structural jamming and tension puckering. The reason for feed puckering is differing friction conditions during fabric feeding (Fig. 6.11). In the sewing machine, it occurs under the following conditions:

a) high friction between upper fabric and presser foot;
b) low friction between upper and lower fabric;
c) high friction between feed dog and lower fabric.[23]

Feed puckering can be prevented by various measures, for example, a Teflon-coated presser foot can decrease friction between upper fabric and presser foot.

Structural jamming is usually caused at the insertion of the sewing thread in the fabric, thus the thread displaces the yarns near the stitch hole. The resulting tension leads to fabric surface unevenness. The higher the fabric count and the thicker the sewing thread, the higher the chance of puckering. One of the ways to avoid this type of puckering is to use the needles with cutting points. The other well-known way to reduce this distortion is to specify seaming at an angle of 15° to the warp direction.

The third factor – tension puckering – is limited to seams stitched with very elastic sewing threads and high thread tension. A thread that has been inserted with high tension will pucker after sewing. In this case, variations of humidity and temperature also play a significant role.[41]

For a long time, the quality problems related to seam puckering were of great interest not only to producers, but also to researchers. A significant number of investigations have been done in this area, starting with the analysis of the influence of sewing machine parameters on seam pucker[42,43] and ending with attempts to simulate this phenomenon. The simulations of Inui and Yamanaka[44] were based on a computer program which predicts the shape of a virtual fabric with an assumption that the relationship between the stress and the strain of the virtual fabric is linear. The simulations predicted the shape in respect to material puckering and mechanical properties of the fabric. Results obtained have shown good

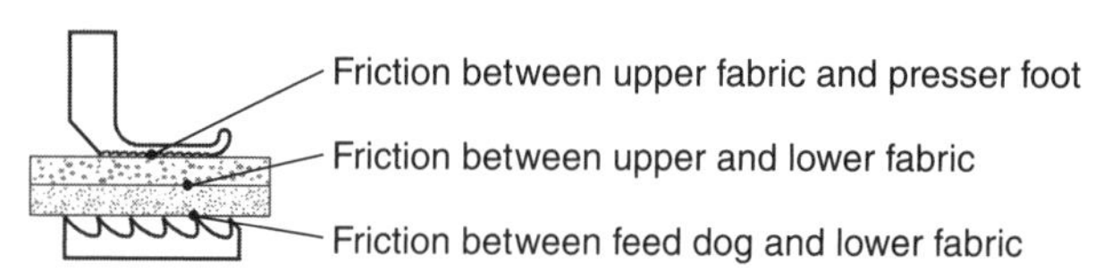

6.11 The conditions for the appearance of feed puckering.

correspondence between theoretical simulations and experimental investigations.

Two of the most troublesome sewability problems in garment manufacture, seam pucker and sewing damage, were considered by Stylios,[45] who has developed intelligent computer-based systems to forecast those problems. The system for the prognosis of seam pucker is based on a mathematical model developed from extensive sewing experiments and on the measurement of fabric thickness and flexural rigidity. The sewing damage system is based on instrumentation and interfacing of an industrial sewing machine with a computer, for real-time dynamic measurements of sewing data.

The problems of seam pucker assessment with respect to fabric characteristics and parameters of a sewing machine have been analyzed by Juciene and Dobilaite.[46] They studied the influence of rotational frequency of the main shaft and pressing force on seam pucker and found that seam pucker sharpness increases with increasing rotational frequency of the main shaft and decreases with increase of the pressing force.

Seam puckering can appear not only on traditional textile materials joined by thread stitching, but also on coated and laminated materials connected by stitchless joints, e.g. sealed. Upholstery materials are the type of fabrics treated by such finishing techniques. Some research has been carried out to compare the behaviour of stitched and sealed seams on breathable waterproof fabrics.[47] It was discovered that the pucker depended on the type of finishing, e.g. for laminated materials it was much higher than for coated ones. The conclusion was made that two factors determine puckering appearance on breathable waterproof fabrics: conditions of fabric transportation and structural jamming. As was discussed above, the first is caused by differential feeding when two pieces of fabric are fed into the gap between a press foot and needle plate and the second is related to sewing thread insertion conditions.

6.3.4 Chain stitch, overedge and specialized sewing machines

Lockstitch and drop feed sewing machines with the added feature of a swinging (zig-zag) needle can be used for embroidery and decoration of soft furnishing because they produce plain and decorative seams. Their application in upholstery is very small, e.g. for curtains and bed-covers. Long-arm sewing machines are essential for fluted and quilted panels. These machines are computer fed and of lockstitch type; their main feature is a long machine arm; and they are classed as heavy duty. Compared to a standard-sized lockstitch machine which has a space from 305 mm to 380 mm, the long-arm machine can have up to 815 mm.[2]

The application of single and two-thread chain-stitch machines in upholstery production is limited, but they are used much more for automotive upholstery. The overedge machines are increasingly used for upholstery work. They can be two, three, or four thread, and single or two needle. The demand for detachable and loose covers has increased the need for seam edges to be overlocked. Stitch formation in this case is more complicated. It is still basically a chain stitch, a second needle providing the safety stitch alongside the overlocking one.

The motor industry is the biggest user of quilting machines for the production of car interior upholstery. These machines can also produce flat sewn panels ready for insertion into a mattress or a chair. Quilters can be single or multi needle. Such specialized equipment as profile seamers sew together an almost unlimited range of sub-assemblies. The operation is carried out using an autojig system, which enables the machines to stitch and to trim one or two plies of fabric to a predetermined shape. Other specialized machines are automatic straight seamers and continuous zippers. The first are equipped with synchronized puller feeds on top and bottom, a double chain-stitch head and thread cutters. They are used where long, straight seams are required, e.g. lengthening or widening of cloth panels or pre-sewing before shaping. The second – continuous zippers – are usually two or four needle and are designed for the fast production of zipped components and continuous zipped borders.

Button making, insertion and fixing during the 1970s and into the 1980s, was a design feature of mass produced upholstery for domestic chairs and sofas to be buttoned in the seat, back and arm areas.[2] At present, fashion demands a less fussy appearance but buttoning – functional and decorative – will remain a part of the upholstery process.

6.4 Future developments and applications

New high-tech fibres and sophisticated material constructions have been developed and are increasingly used for the production of modern interiors textiles, e.g. Spectro Coating Corp. company has introduced a new line of high-performance upholstery fabrics for commercial, hospitality and residential furniture, and for automotive, recreational vehicle and marine interiors. This upholstery fabric is called *Primo 150* and features a nanotechnology-based, stain-resistant finish.[48] The appearance of such new and advanced materials with high-performance properties inspires users to replace classic sewing processes, e.g. sewn seams, with advanced techniques such as stitchless ultrasonic, thermal or high-frequency joints.[49] These methods refer to joining substrates without sewing thread and they can be obtained by fusing, adhesive seaming and welding. Welding is achieved by ultrasonic vibrations, radio frequency heating or infrared beams applied to the ply to be welded. This is possible only for 100% thermoplastic materials or for those having less than 35% of natural fibres or other non-thermoplastic fibres blended with the thermoplastic material.

One of the biggest producers of such equipment is PFAFF Industrie Maschinen company which produces ultrasonic seam welding and seam sealing machines, as well as continuous fusing machines and pressers.[50] The ultrasonic welding machine *Seamsonic 8310* (winner of the 2006 IMB innovation award) is designed for most thermoplastic sheets, laminates, non-wovens and textiles with a high proportion (70% or more) of weldable polymer. Application examples are medical mattresses, pillows, shower curtains, covers for cars and sunshades. Sealing takes place as a result of ultrasonic oscillations of a sonotrode and an anvil wheel. Process reliability is quaranteed through monitoring of welding parameters. User can apply the anvil wheel with a specific seam look. Seam width can vary between 2 and 10 mm, maximal speed 0.6–13.6 m/min, and ultrasonic frequency 35 kHz.

An example of a plastics heat sealing machine is the PFAFF *Weldfing 8309*, produced as a post type, hot-wedge model. It is considered to supply a safe heat-sealing method for the future, which enables handling of such modern materials as PE, PP and not just PVC and PU. The processing of heavy, large-surface materials with this machine becomes simple. Heat sealing temperature can vary from 20 °C up to 450 °C, sealing speed is 0–20 m/min, seam width can be 10, 20, 30, 40 and 50 mm, material thickness from 0.2 mm. An electronic control unit makes it possible to adapt the speed after sealing has started. Another example is the PFAFF *Weldchampion 8304* universal plastics sealing machine, which can operate both as a hot-wedge and/or a hot air model. In the case of heat-sealing with hot air, the temperature can reach 450 °C, while in the case of heat sealing with hot wedge, up to 600 °C.[50]

The strength and performance properties of welded seams are important objects for the investigations. Vujasinovic *et al.*[51] applied a PFAFF *Seamsonic 8310* ultrasonic welding machine with a frequency of 30 kHz for seam strength investigations. Titanic sonotrodes, 104 mm in diameter, were used. Their maximal weld width was 10 mm. Four anvil wheels with different engraving (point, zigzag, one-line and three lines) were applied for ultrasonic seam welding. Comparative analysis of seam strength was performed with samples prepared by a special sewing machine applying a threefold zigzag stitch of polyester thread. The authors concluded that ultrasonic welding, together with the selection of optimal parameters, e.g. speed and amplitude, could successfully replace classical sewing operations because the strength of obtained bond was, as a rule, higher than the strength of the stitched seam. Also, the welded bond was approximately 2/3 thinner than the stitched seam. It was less rough and, in the case of optimally selected welding conditions, it was completely air impermeable. The same results were obtained for fabrics with a special waterproof finish.[47]

Other literature sources note that welded seams are characteristically stiff and harsh, and a number of claims are made that welded seams are weaker than sewn ones and that welding can decrease fabric strength.[8] Limited research has been done and little is published concerning the performance of welded seams compared with sewn seams, and with the effects of joining process parameters upon fabric strength. Thus, existing results sometimes are contradictory. This proves the need for further and thorough investigations of new joining technologies and analysis of seam performance properties, especially those concerning the new generation of interior textiles.

Further developments in fabric joining technologies are also associated with laser welding, which provides high quality seams with both good aesthetic appearance and performance properties, and offers an excellent opportunity to increase production automation. This process is non-contact and is applied only from one side of the joint. Laser technology can significantly improve seam barrier properties, reliability and endurance by replacing the two-stage process of stitching and seam taping with welding. This method was practically applied for a chair supplied by UK-based Knightsbridge Furniture Productions Ltd, which featured a waterproof cover of PU-coated polyester fabric.[52] Hermetically sealed seams, which decrease bacterial penetration to the seat interior and reduce the risk of contamination and transfer of infection, were required by the healthcare industry. The covers were successfully welded by laser technologies. Afterwards, durability tests simulating a year of service was performed on the chair with no significant damage and the seams passed hydrostatic pressure leak tests. Similar practical experiments were also performed with UK Silentnight Beds company, which selected three separate bed manufacturing operations: attaching the mattress information label, producing a welded joint in the mattress side-panel border in quilted fabric, and welding the fabric on the PVC-coated wooden drawer front of a divan. All of these applications were successfully welded and subjected to industry standard testing with no visible deterioration of the seams.

Thus, laser welding of fabrics can lead to greater automation, increased productivity and improved quality, offering manufacturers a competitive advantage. Additionally, the process reduces noise levels and injuries in the workplace. Further developments in fabric joining technologies promise great benefits in terms of process speed, automation and quality improvements of interior textile products.

6.5 References

1. CLIVE, E. *Encyclopedia of Furniture Materials, Trades and Techniques.* Aldershot: Asjgate Publishing Limited, 2000.
2. JAMES, D. *Upholstery: A Complete Course.* Lewes: GMC Publications, 2001.
3. MILLER, J. In Pursuit of Quality – The Volume Control for Fabric Goods is Getting out of Hand. *International Journal of Consumer Studies*, 22(4), 2007, pp.191–198.
4. ISO 4915:1991 *Textiles – Stitch types – Classification and Terminology.*
5. JUCIENE, M., VOBOLIS, J. Survey of the Lockstitch Quality Effecting Upholstered Furniture. In: *Proceedings of the III International Scientific Conference Light Industry – Fibrous Materials*, Radom, 17–18th of November 2005. pp.328–333.
6. AMANN, *Sewing Threads.* Bönnigheim: Amann & Söhne GmbH & Co., pp.40–44.
7. MUKHOPADHYAY, A. Relative Performance of Lockstitch and Chain-stitch at the Seat Seam of Military Trousers. *Journal of Engineered Fibres and Fabrics*, 3(1), 2008, pp.21–24.
8. LAING, R.M., WEBSTER, J. *Stitches and Seams.* Manchester: The Textile Institute, 1998.
9. ISO 4916:1991 *Textiles – Seam types – Classification and Terminology.*
10. TARTILAITE-JUCIENE, M., VOBOLIS, J. The Effect of Sewing Fabric Parameters upon the Seam Quality. *Materials Science. Medziagotyra*, 8(1), 2002, pp.116–119.
11. JUCIENE, M., VOBOLIS, J. Correlation Between the Seam Stitch Length of the Sewing Garment and Friction Forces. *Materials Science. Medziagotyra*, 13(1), 2007, pp.74–77.
12. SUNDARESAN, G., HARI, P.K., SALHOTRA, K.R. Strength Reduction of Sewing Threads during High Speed Sewing on an Industrial Lockstitch Machine: Part I – Mechanism of Thread Strength Reduction. *International Journal of Clothing Science and Technology*, 9(5), 1997, pp.334–345.
13. ZUNIC-LOJEN, D., GERSAK, J. Determination of the Sewing Thread Friction Coefficient. *International Journal of Clothing Science and Technology*, 15(3/4), 2003, pp.241–249.
14. MASTEIKAITĖ, V. Seam Stiffness Evaluation. In: *Proceedings of 2nd International Conference Innovation and Modelling of Clothing Engineering Processes*, Maribor 8–10th, October 1997, pp.114–121.
15. UJEVIC, D., KOVACEVIC, S. Impact of the Seam on the Properties of Technical and Nonwoven Textiles for Making Car Seat Coverings. *International Nonwovens Journal*, 13(1), 2004, pp.33–41.
16. BEIER, U., FISCHER, F., SANDLER JAN, K.W., ALTSTÄDT, V., WEIMER, CH., BUCHS, W. Mechanical Performance of Carbon Fibre Reinforced Composites Based on Stitched Preforms. *Composites Part A: Applied Science and Manufacturing*, 38(7), 2007, pp.1655–1663.

17. WEIMER, C., MITSCHANG, P. Aspects of the Stitch Formation Process on the Quality of Sewn Multi-textile-preforms. *Composites Part A: Applied Science and Manufacturing*, 32(10), 2001, pp.1477–1484.
18. MOURITZ, A.P., LEONG, K.H., HERSZBERG, I. A Review of the Effect of Stitching on the in-plane Mechanical Properties of Fibre-Reinforced Polymer Composites. *Composites Part A: Applied Science and Manufacturing*, 28(12), 1997, pp.979–991.
19. EBERLE, H., HERMELING, H., HORNBERGER, M., MENZER, D., RING, W. *Clothing Technology from Fibre to Fashion*. Nourney: Verlaf Europa-Lehrmittel, 2002.
20. LOMOV, L. A Predictive Model for the Penetration Force of a Woven Fabric by a Needle. *International Journal of Clothing Science and Technology*, 10(2), 1998, pp.91–103.
21. STYLOS, G., XU, Y.M. An Investigation of the Penetration Force Profile of the Sewing Machine Needle Point. *Journal of the Textile Institute*, 86(1), 1995, pp.148–163.
22. JANKAUSKAITE, V. STRAZDIENE, E., LAUKAITIENE, A. Stress Distribution in Polymeric Film Laminated Leather under Biaxial Loading. *Proceedings of the Estonian Academy of Sciences. Engineering*, 12(2), 2006, pp.111–124.
23. RHEIN NADEL. Aachen: Rhein-Nadel Maschinennadel GmbH, 1995.
24. KHAN, R.A., HERSH, S.P., GRADY, P.L. Simulation of Needle–Fabric Interactions in Sewing Operations. *Textile Research Journal*, 40(6), 1970, pp.489–498.
25. LIASI, E., DU, R., SIMON, D., BUJAS-DIMITREJEVIC, S., LIBURDI, F. An Experimental Study of Needle Heating in Sewing Heavy Materials using Infrared Radiometry. *International Journal of Clothing Science and Technology*, 11(5), 1999, pp.300–314.
26. LI, Q., LIASI, E., HUI-JUN, Z., DU, R. A Study on the Needle Heating in Heavy Industrial Sewing. Part 1: Analytical Models. *International Journal of Clothing Science and Technology*, 13(2), 2001, pp.87–105.
27. LI, Q., LIASI, E., SIMON, D., DU, R. A Study on the Needle Heating in Heavy Industrial Sewing: Part 2: Finite Element Analysis and Experiment Verification. *International Journal of Clothing Science and Technology*, 13(5), 2001, pp.351–367.
28. SUNDARESAN, G., SALHOTRA, K.R., HARI, P.K. Strength Reduction in Sewing Threads during High Speed Sewing in Industrial Lockstitch Machine: Part II: Effect of Thread and Fabric Properties. *International Journal of Clothing Science and Technology*, 10(1), 1998, pp.64–79.
29. ZILIENE, L., BALTRUSAITIS, J. Investigation of Sewing Thread Damage during Sewing Process. *Material Science. Medziagotyra*, 6(2), 2000, pp.104–108.
30. MICHAEL H.G., VINGILIO, D.R., MACK, E.R. Sewing Needle Temperature. Part III: The Effects of Sewing Conditions. *Textile Research Journal*, 43(11), 1973, pp.651–656.
31. RUDOLF, A., GERSAK, J., UJHELYIOV, A., SFILIGOJ SMOLE, M. Study of PES sewing thread properties. *Fibres and Polymers*, 8(2), 2007, pp.212–217.
32. ZUNIC-LOJEN, D., GOTLIH, K. Computer Simulation of Needle and Take-up Lever Mechanism Using the ADAMS Software Package. *Fibres & Textiles in Eastern Europe*, 11(4), 2003, pp.39–44.
33. KAMATA, Y., KINOSHITA, R., ISHIKAWA, S., FUJISAKI, K. Disengagement of Needle Thread from Rotating Hook, Effects of its Timing on Tightening Tension,

Industrial Single-Needle Lockstitch Sewing Machine. *Journal of the Textile Machinery Society of Japan*, 30(2), 1984, pp.40–49.

34. FERREIRA, F.B.N., HARLOCK, S.C., GROSBERG, P. A Study of Thread Tensions on a Lockstitch Sewing Machine. Part I. *International Journal of Clothing Science and Technology*, 6(1), 1994, pp.14–19.
35. OSAWA, M., SAITO, H. Measurement of Feed Dog Motion and Dynamic Torque in Sewing Machines by Means of an Optical Lever Device. *Journal of the Textile Machinery Society of Japan*, 29(3), 1983, pp.66–71.
36. ZAJACZKOWSKI, J. Effect of Belt Extensibility on Variation of the Relative Position of a Needle and a Hook in a Sewing Machine. *International Journal of Clothing Science and Technology*, 12(5), 2000, pp.303–310.
37. STYLIOS, G., SOTOMI, O.J. A Neuro-fuzzy Control System for Intelligent Sewing Machines. In: *Proceedings of 2nd International Conference on Intelligent Systems Engineering*, 5–9th of September 1994. pp.241–246.
38. ROBAK, D.A. Model of Fabric Transport in a Sewing Machine Using a Feed Dog Covered with a Supple Material of Increased Friction. *Fibres & Textiles in Eastern Europe*, 10(3), 2002, pp.58–62.
39. KIKUKO, A., KOZO, S. A Study on Easing by the Bottom and Variable Top Feed Sewing Machine. Influence of Sewing Condition and Fabric Properties on Easing. *Fibre*, 58(6), 2002, pp.216–223.
40. GOTTSCHALK, T., SELIGER, G. Automated Sewing of Textiles with Different Contours. *CIRP Annals – Manufacturing Technology*, 45(1), 1996, pp.23–26.
41. DOBILAITE, V., JUCIENE, M. The Influence of Mechanical Properties of Sewing Threads on Seam Pucker. *International Journal of Clothing Science and Technology*, 18(5), 2006, pp.335–345.
42. MORI, M., NIWA, M., KAWABATA, S. Effect of Thread Tension on Seam Puckering. *Sen'i Gakkaishi*, 53(6), 1997, pp.217–225.
43. DOBILAITE, V., JUCIENE, M. The Influence of Sewing Machine Parameters on Seam Pucker. *Tekstil*, 56(5), 2007, pp.286–292.
44. INUI, S., YAMANAKA, T. Seam Pucker Simulation. *International Journal of Clothing Science and Technology*, 10(2), 1998, pp.128–142.
45. STYLIOS, G. Prognosis of Sewability Problems in Garment Manufacture Using Computer-based Technology. In *Proceedings of IEEE International Conference Systems Engineering*, 1990. pp.371–373.
46. JUCIENE, M., DOBILAITE, V. Seam Pucker Indicators and their Dependence upon the Parameters of a Sewing Machine. *International Journal of Clothing Science and Technology*, 20(4), 2008, pp.231–239.
47. JEONG, W.Y., AN, S.K. Seam Characteristics of Breathable Waterproof Fabrics with Various Finishing Methods. *Fibres and Polymers*, 4(2), 2003, pp.71–76.
48. http://www.textileworld.com/Articles/2008/May_2008/Textile_News/Spectro_Introduces_New_Primo_150_Upholstery_Fabrics.html, 03 06 2008 [accessed 17 12 2008].
49. UJEVIC, D., ROGALE, D., KARABEGOVIC, I. Achievements and Trends of the Development of Sewing Machines Exhibited at the IMB 2003. *Tekstil*, 53(5), 2004, pp.245–255.
50. PFAFF INDUSTRIE MASCHINEN COMPANY, 2008. www.pfaff-industrial.com [accessed 17 12 2008].

51. VUJASINOVIC, E., JANKOVIC, Z. Investigation of the Strength of Ultrasonically Welded Sails. *International Journal of Clothing Science and Technology*, 19(3/4), 2007, pp.204–214.
52. IAN, J. Improving productivity and quality with laser seaming of fabrics. *Technical Textiles International: TTI* [online], May 2005. Available from: http://findarticles.com/p/articles/mi_qa5405/is_200505/ai_n21373378/pg_1?tag=artBody;co] [accessed 17 12 2008].

7 Environmental issues in interior textiles

L. TUCKER, Virginia Tech, USA

Abstract: The chapter begins by discussing some of the key environmental issues related to interior textile production. It then reviews trends in environmentally responsible textile design and production followed by an overview of some textile certification programs. The chapter concludes with advances in medical textiles, bio textiles and e-textiles.

Key words: e-textiles, bio-textiles, green chemistry, textile certification systems.

7.1 Introduction

The primary focus of this chapter is upon recent environmental perspectives on interior textiles and related new developments that have revolutionized the industry during the past decade. A renewed interest and focus on sustainable design and the resulting increased governmental regulation of the textile industry around the world has resulted in major changes to the production of textiles. Combined with recent technological innovations in textiles, new possibilities in the industry abound.

7.2 Key environmental issues

Environmental concerns include waste, reliance on man-made, petro-based fibers, and other hazardous materials. A major environmental concern related to textiles is the dye effluent produced from textile factories.

7.2.1 Waste

Traditionally, waste from textile plants has been considered toxic. When the trimmings from their textiles were classified as hazardous waste by the Swiss Government, Rohner AG met with Susan Lyons of Design Tex and William McDonough and Michael Braungart of McDonough Braungart Design Chemistry (MBDC). Facing more stringent regulations, they decided that the only way to change this was to radically alter the way that the textiles were made. A major element of this process was to find non-toxic dyes. Of the 60 companies MBDC invited to participate in the dye testing process only one agreed – Ciba-Geigy. Of the 1600 dye formulas provided by this company and reviewed by Braungart and McDonough, only 16 were deemed toxin free. Using these 16 dyes and new manufacturing methods, the trimmings can now be composted as yard mulch and are said to be safe enough to ingest. Remarkably, once DesignTex had reinvented their manufacturing processes for these new textiles, they made the technology available to other manufacturers. Only Carnegie Fabrics responded by signing an eco-cooperation agreement with DesignTex. Thus the 'Climatex' fabric resulting from the Rohner-MBDC-DesignTex collaboration is now used by both DesignTex and Carnegie Fabrics (Bonda, 2002).

7.2.2 Recycling

One of the most important aspects of the redesign process was the idea that 'waste equals food,' an idea presented by McDonough and Braungart (2002). The ultimate goal was to eliminate the concept of waste. In their book *Cradle to Cradle*, McDonough and Braungart (2002) refer to design as the first signal of human intention. They proceed to outline a process by which the design of anything can be transformed from a 'cradle to grave' process to a 'cradle to cradle' process. The latter process requires that biological nutrients remain separated from technical nutrients, which allows for the infinite recycling of each. Since this time, Rohner AG and DesignTex have led a revolution in the production of interior textiles (Bonda, 2002).

7.2.3 Water pollution

One of the key concerns associated with textile production is water pollution. The wastewater (or effluent) from the textile industry introduces a variety of bi-products into the water supply – both organic and inorganic. These include, but are not limited to, ions, metals, solvents, nutrients, detergents and phthalates, and result in changes to the water bodies including color, temperature and odor. A recent study conducted by Gomez *et al.* (2007) has demonstrated significant changes in habitat and water quality downstream from textile manufacturing plants in Argentina.

Much research has been conducted on how to reduce the impact of toxins from textile effluent. A recent study performed by Brazilian scientists demonstrates that the introduction of aerobic processes caused the degradation of industrial textile effluents and was therefore considered 'environmentally friendly' (Ulson de Souza *et al.*, 2007). Thus the introduction of aerobic processes is a possible solution to pollution.

Rohner AG and Design Tex took an entirely different approach to effluent through the elimination of water pollution as a bi-product of textile manufacturing. Remarkably, the effluent now being produced from the Rohner AG production facility is cleaner than the water going in to the production process in the first place (Bonda, 2002).

7.2.4 Dyes

Greaves and coauthors' (1999) comprehensive review of the literature concerning the removal of dyes from textile effluent contains six papers written between 1969 and 1999. The researchers sought to understand why, after so many years of investigation into the removal of dyes from effluent and several suggested solutions, many effluent works still did not use the new technology. The methods that various researchers had proposed during this time include: membranes, flocculation, electrochemical oxidation, coagulation, and aerobic and anaerobic biological treatments (Churchley *et al.*, 2005). What the researchers discovered was that each of the studies followed different testing protocols, making them hard to compare, and this sometimes led to contradictory results. They concluded by saying that most of the methods of elimination that have been proposed during the preceding 30 years were largely ineffective, thus explaining their lack of industrial integration. However, the Brazilian study mentioned previously (Ulton deSouza *et al.*, 2007) showed some success with aerobic biological treatments.

In 2002, Hong Kong researchers Kwok *et al.* created and demonstrated a protocol that could be used to accurately predict the degree of pollution in after-dying effluent for specific dyes. Although the model was tested using only single dyes, it does provide a standard protocol for use by future studies.

As environmental regulations of the textile industry become more stringent around the world, some of the recent research has focused on the environmental properties associated with specific dyes. As an example, Edwards and Freeman (2005) have studied formazan dyes. Previous research had already demonstrated other technical strengths of these dye types, including wet fastness and photostability; however, the researchers in this study wanted to test the effluent toxicity resulting from these dyes. They discovered that the iron complex formazan dyes had a lower toxicity than the iron complex azo dyes.

As the most pollution intensive part of the textile industry, the reduction of toxicity in dyes as well as the successful treatment of effluent are the two primary goals for research as it relates to the environment and textiles. Many efforts are being made on behalf of the textile industry to meet increasingly stringent governmental standards for effluent treatment, including water decoloration through absorption of dyes. Recent research efforts have focused on microbial decoloration strategies (Kwok *et al.*, 2002). A reduction in the use of hazardous chemicals has led to the development of several environmentally-friendly enzymes to assist in the process. These include lacasse, catalase, alpha-amylase, and pectinase. (Teli, 2008). The use of biotechnology to treat textiles and their byproducts continues to grow.

7.3 Textile certification programs

With the rise of new technologies and interest in sustainable textiles, many textile certification programs have been developed around the world.

7.3.1 Commercial Furnishings Fabrics Sustainability Assessment

Work is underway on a uniform sustainable textile standard called the 'Commercial Furnishings Fabrics Sustainability Assessment'. It is being written by a group comprising GreenBlue (another of the McDonough–Braungart companies), the National Science Foundation (NSF), academia and industry members. The work has been ongoing for the past couple of years but has not yet (early 2009) been released to the public. The certification will be based on a points system where points are assigned in the compliance categories listed in Table 7.1. The latest draft of the document can be found on the standards list of the NSF site.

Table 7.1 Proposed points breakdown of Commercial Furnishings Fabrics Sustainability Assessment

Title	Prerequisite credit points	Optional credit points available (maximum)
Fiber Sourcing	0	20
Safety of Materials	8.4	21.6
Water Conservation	1.6	5.2
Water Quality	4.3	0.9
Energy	2	10
Air Quality	2	0
Reduction, recycling, and re-use in manufacturing and end of use	3	9
Social Accountability	1	11
Total	22.3	77.7

Adapted from March 2008 Draft of NSF Document 336.

7.3.2 OEKO-Tex Standard 100 (International Oeko-Tex Association)

Originally started in the 1990s, the OEKO-Tex Standard 100 provides testing and certification for the raw materials used in the textile industry. The desire was to provide non-toxic textiles that presented no harm to the end-user. The OEKO-Tex 100 label ensures that the textile has met a globally uniform and scientifically-based set of criteria, determined by independent testing institutes on an annual basis. The OEKO-Tex Standard 1000 complements the previous standard, adding to it the use of environmentally-friendly facilities and processes coupled with the prohibition of child labor.

7.3.3 ISO 14001 (International Standards Organization)

The ISO 14001 is a part of the ISO 14000 family of standards. These standards deal with the environmental management systems of a company. The standard provides guidelines without specific restrictions in an attempt to create a holistic approach to environmental concerns within a company.

7.3.4 EPEA (Environmental Protection Encouragement Agency)

EPEA was founded in Germany by Dr Michael Braungart in 1987. Like GreenBlue in the US, EPEA works to design materials and processes that meet cradle-to-cradle protocol. EPEA, Material ConneXion, and MBDC have recently entered into an agreement regarding sustainable material solutions. Like MBDC, EPEA works with industry to identify problems and design solutions to meet cradle-to-cradle protocol. EPEA provides cradle-to-cradle certification for compliant materials and products. Material ConneXion provides an ongoing list of products and materials that meet the certification as well as other cutting edge materials.

7.3.5 SMART Sustainable Textile Standard (Institute for Market Transformation to Sustainability, USA)

The SMART Sustainable Textile Standard 2.0 was adopted in 2004. As a voluntary program, the goal of the standard is fivefold: (i) increase the demand for sustainable textiles and thus their economic value, (ii) help specifiers understand sustainable textile characteristics, (iii) create a consensus based standard, (iv) educate people at all steps of the textile process, and (v) encourage competition to enhance sustainability in the textile market. The SMART Standard covers the environmental impact of the entire life-cycle of a textile and its supply chain. Five categories are used to evaluate the textile: (i) Safety for Public Health and Environment, (ii) Renewable Energy and Energy Efficiency, (iii) Material, Biobased or Recycled, (iv) Facility and Company-based achievements and Social Indicators, and (v) Reclamation, Sustainable Re-use and End-of-life Management.

Safety of chemical and materials inputs

Under the SMART Standard, chemical input is regulated consistent with the Stockholm Convention Persistent Organic Chemicals list. These POPs (Persistent Organic Pollutants) remain in the human system for extended periods of time and are considered toxic. The Stockholm Convention covers twelve POPs: Aldrin, Chlordane, DDT, Dieldrin, Dioxins, Endrin, Furans, Heptachlor, HCB, Mirex, Polychlorinated biphenyls (PCBs), and Toxaphene. Many of these toxins are insecticides and are used around crops, thus becoming a serious issue in the use of natural or biobased fibers. The chemical criteria apply to both inputs and emissions from textile manufacturing.

Energy efficiency and mix

The SMART standard addresses both renewable energy use and overall energy efficiency in the textile production process.

Recycling and actual reclamation

Points are assigned within the SMART Textile Standard based on the company's reclamation efforts, the best leading to silver, gold, and platinum sustainable achievement ratings.

Social equity for workers

Textiles are produced around the world and, as a result, strict measures need to be taken to ensure the social equity of the workers involved.

7.4 Types of sustainable and recycled textiles used for interiors

7.4.1 Cradle to cradle textiles

As mentioned previously, the original work done with Rohner AG and DesignTex led to the development of 'Climatex', a cradle-to-cradle compliant textile. The key considerations to meet this protocol include the separation of biological and technical nutrients, the elimination of waste, and the notion of working within nature's systems. Ultimately, the products do no harm to the natural environment and can be infinitely recycled.

Biological nutrients

A biological nutrient is one found in nature with no man-made content. It is a biodegradable material posing no immediate or eventual hazard to living systems that can be used for human purposes and can safely return to the environment to feed environmental processes, e.g. cotton, linen, sisal.

Technical nutrients

A technical nutrient is one developed with mankind's intervention e.g. petro-based textiles. The material remains in a closed-loop system of manufacture, reuse, and recovery (the technical metabolism), maintaining its value through many product life cycles.

7.4.2 Recycled textiles

Recycled textiles are made from either post-consumer or post-industrial materials. Recycled textiles made from plastic bottles, for example, are post-consumer. Polyester (PET) plastics are made from petroleum and can be recycled. An excellent example of a product that uses both post-industrial and post-consumer polyester is Interface's *Terratex* classified polyesters. Until recently, all polyester has contained traces of antimony (a metal), used as a catalyst in production. Some European textile manufacturers have developed technologies that are antimony-free. Two examples include Devan Chemicals in Belgium and Zimmer in Germany (Otto, 2007).

7.4.3 Recyclable textiles

Since McDonough's work with DesignTex, the notion of recyclable textiles has become commonplace. Textiles where biological nutrients and technical nutrients are not mixed (such as wool/nylon blends), can be theoretically recycled. Of greater importance at the time of this writing (2009) is the need for an established collection system for used textiles. Like the carpet industry, which has also produced recyclable goods, the textile industry also needs to address recapture of these materials to avert them from the landfills of the world.

7.4.4 Bio-textiles

A bio-textile is one that relies on a biological source, such as wheat, corn, rice, bamboo and soy. An emerging market of importance is the bio-textiles industry. A recent article (Huda *et al.*, 2007) claims the following about the US textile industry: 'New biofibers derived from the byproducts of the agriculture and biofuel industries and other renewable sources will be critical for the future competitiveness and sustainability of the textile industry in this country.' Within the article, the authors discuss natural cellulose fibers derived from rice, corn, sugarcane, pineapple, in addition to protein fibers from wheat and soybeans. The use of bamboo fibers (and soy-based foams) are becoming commonplace in the market. Although the technology to produce these fibers has been in place in some instances since Ford first regenerated proteins from soybeans, the cheaper prices of other natural fibers and eventually man-made fibers dominated the industry. As fuel costs continue to rise, a renewed interest in some of these biobased fibers has resurfaced.

7.5 Applications/examples

Many textile companies are now working with and marketing sustainable or 'green' textiles. Most major manufacturers have participated but not all interventions are equal. Most consider the cradle-to-cradle certification process to be the most stringent third-party testing in the marketplace. Companies that currently produce cradle-to-cradle compliant textiles for interior applications include Rohner Textil AG, Craftex, DesignTex (Climatex), Spectrum Yarns, Pendelton Woolen Mills, Victor Innovatex, Weave Corporation, Miliken and Company (180 degree walls) and Carnegie (Climatex, Surface IQ, and Xorel). Several other organizations and testing agencies also operate in the sustainability trade. These include Green Seal, Green Guard, The Carpet and Rug Institute GreenLabel Program, Scientific Certification Systems, GreenSpec and GreenPrints, to name a few of the more commonly occurring ones.

7.6 Future trends

China ranks first in the world with its production capacity for woolen fabrics, silk fabrics, chemical fibers and cotton textiles. As such, research programs in China will likely have a significant impact on the future of textiles. China has produced naturally-colored cottons and silks through genetic modification. The Chinese are also working with genetic engineering of silk worms to produce silk which wrinkles less and holds color better. With heightened environmental controls, the textile industry in China has had to look to more environmentally-friendly techniques for textile preparation. Bio-degumming of fibers using microbes has produced a series of new technologies in Chinese textiles within the past few years. China alone has been responsible for some 0.65 billion tons per annum of wastewater from the textile process (Chen *et al.*, 2006). Textile effluent is now being treated using additional anaerobic–aerobic processes. Textile biotechnology has become a major focus of the 11th Five Year Plan for China (2006–2010).

7.6.1 Current research directions

Interface (2008) has been conducting extensive research with regard to sustainable textiles with the 'Mission Zero' goal of becoming a company with no environmental impact by the year 2020. As part of their research agenda, they have created recycled products, recyclable products, biobased fibers, and other sustainable textile initiatives. As a part of this, they have conducted life-cycle assessments on all their products. One discovery from this process is that their Terratex protocol fabrics, which rely on recycled polyester, have a much lower environmental impact than their fabrics using new polyester. By recycling polyester, the company claims to have saved 484,150 barrels of oil and 74 million gallons of water between 1996–2001.

In addition to research being done within textile companies, independent researchers at academic institutions across the world have been conducting a broad range of textile research on a variety of issues including natural fiber alternatives, the use of feathers for fibers, the eco-efficiency of certain textiles, and consumer attitudes towards textile recycling.

7.6.2 Scent-infused textiles

A team at Philadelphia University explored the integration of scents into polymer fibers (Pierce, 2005) with the goal of enhancing consumer experiences. Funded by the National Textile Center, this research explored the competitive advantage of adding pleasant smells to textiles to increase their marketability. Additional research funded by the National Textile Center has explored odor reduction using bioorganisms or reactive chemical species.

7.6.3 Smart fabrics (electronic textiles, e-textiles, intelligent textiles)

E-textiles have electronic components woven directly into them including electronics and interconnections. Research on electronic/intelligent textiles is being done around the world to explore the many possible applications of variations on this product-type.

7.6.4 Fabrics that track medical data

BIOTEX, a European project, has succeeded in weaving miniature biosensors into a textile patch, allowing the textile to track and analyze body fluids in addition to body temperature. Jean Luprano, Swiss Center for Electronics and Microtechnology, heads the BIOTEX Project. Similar inventions include textiles that can track heart rate and respiratory function.

The incorporation of electronic components into textiles, if adopted on a large scale, will have widespread ramifications on their recyclability.

Carbon nanotubes

CSIRO Textile and Fibre Technology (CTFT) has created fibers spun from carbon nanotubes which are able to conduct electricity and heat. (Science News, 2008) Although most applications to date have involved wearable textiles, the application to interior textiles is likely not far behind.

7.7 Sources of further information and advice

Many advances in textile design are occurring outside of the traditional channels and include electronics and scientific research. While many of these advances have been developed for use in the apparel industry, their impact on interiors has yet to be explored. Some useful sources of information and advice are as follows:

- Association for Contract Textiles (ACT) http://www.contracttextiles.org/main.php?view=start
- EPEA (Environmental Protection Encouragement Agency) www.epea.com
- ISO 14001 (International Standards Organization) www.iso.org
- OEKO-TEX Standard 100 www.eco-tex.org
- Sustainable Textile Standard (ANSI) www.greenblue.org and http://standards.nsf.org
- E-Textile Research Lab www.ccm.ece.vt.edu/etextiles/
- Scented Textiles http://www.fitfibers.com/publications.htm
- MBDC http://www.mbdc.com/
- GreenBlue http://www.greenblue.org/
- Rohner AG http://rohnerchem.com/
- DesignTex http://www.designtex.com/

7.8 References

BONDA, P. (2002), 'How Green is my textile?' *Interiors and Sources Magazine*, July–August, 20–25.

CHEN, J., WANG, Q., and DU, G. (2006), 'Research and application of biotechnology in textile industries in China,' *Enzyme and Microbial Technology*, 40 (7), 1651–1655.

CHURCHLEY, J.H., GREAVES, A.J., HUTCHINGS, M.G., PHILLIPS, D.A.S., and TAYLOR, J.A. (2000), A chemometric approach to understanding the bioelimination of anionic, water-soluble dyes by a biomass; Part 3: Direct dyes, *Coloration Technology,* Vol. 116. No. 9, pp 279–284.

EDWARDS, L.C. and FREEMAN, H.S. (2005), 'Synthetic dyes based on environmental considerations. Part 4: Aquatic toxicity of iron-complexed formazan dyes,' *Coloration Technology*, 121 (5), 271–274.

GOMEZ, N., SIERRA, M.V., CORTELEZZI, A., and RODRIGUEZ CAPTUULO, A. (2007), 'Effects of discharges from the textile industry on the biotic integrity of benthetic assemblages,' *Ecotoxicity and Environmental Safety*, 69 (3), 472–479.

GREAVES, A.J., PHILLIPS, D.A.S., and TAYLOR, J.A. (1999), 'Correlation between the bioelimination of anionic dyes by an activated sewage sludge with molecular structure. Part 1: Review,' *Coloration Technology*, 115 (12), 363–365.

HUDA, S., REDDY, N., KARST, D., XU W., YANG, W., and YANG, Y.Q. (2007), 'Nontraditional biofibers for a new textile industry,' *Journal of Biobased Materials and Bioenergy*, 1 (2), 177–190.

Interface 'Terratext' series downloaded from www.terratext.com, March 2008.

KWOK, W.Y., XIN, J.H., and SIN, K.M. (2002), 'Quantitative prediction of the degree of pollution of effluent from reactive dye mixtures' *Coloration Technology*, 118 (4), 174–180.

MCDONOUGH, W. and BRAUNGART, M. (2002), *Cradle to Cradle: Rethinking the Way we Make Things.* New York: North Point Press.

OTTO, B. and BERGER, U. (2007), 'Zimmer unveils eco-friendly heavy metal-free catalyst systems for PET production,' *Fiber Journal*, 16 (3), 62–67.

PIERCE, J. (2005), 'Scent infused textiles to enhance consumer experience.' FIT publications, retrieved May 13, 2008 at http://www.fitfibers.com/publications.htm

SCIENCE NEWS (2008), 'Smart Clothes that track your health,' *Science Daily.* Retrieved from http://www.sciencedaily.com/releases/2008/03/080329121141.htm on May 13, 2008.

TELI, M.D. (2008), 'Textile Coloration Industry in India,' *Coloration Technology*, 124 (1), 1–13.

ULSON DE SOUZA, A.A., BRANDAO, H.L., ZAMPORLINI, I.M., SOARES, H.M., GUELLI, A., and ULSON DE SOUZA, S.M. (2007), 'Application of fluidized bed bioreactor for COD reduction in textile industry effluents,' *Resources, Conservation and Recycling*, 52 (2), 511–521.

8 The role of textiles in indoor environmental pollution: Problems and solutions

L. NUSSBAUMER, South Dakota State University, USA

Abstract: Textiles provide interiors with a sense of softness, they absorb sound, and they can improve aesthetics; however, they can negatively affect indoor air quality (IAQ) due to chemicals used in production, treatments, and more. An illness related to poor indoor air quality is *multiple chemical sensitivity* (MCS). Exposure to various chemicals found within the environment may trigger symptoms of MCS. The major cause of the illness is poor IAQ. This chapter will discuss the concerns for indoor environmental pollution, illness related to poor IAQ, problems with textiles and methods of minimizing pollution through textiles, application and examples, and future trends.

Key words: indoor air quality, multiple chemical sensitivity, microbial pollution, dust mites and animal dander, volatile organic compounds.

8.1 Introduction

Knowledge that outdoor air pollution can damage health is well known. However, evidence has proven that indoor air is more highly polluted and has a greater effect on health than outdoor air. The reason is that people spend about 90 percent of their time indoors within tightly sealed buildings. Thus, many people are at greater risk from health problems because of the increased danger of exposure to pollution indoors than outdoors, and people exposed to indoor air pollutants for longer periods of time are more susceptible to the effects of indoor air pollution (EPA, 1995; Mate *et al.*, 2004; Nussbaumer, 2004; Pilatowicz, 1995).

Since textiles are used within our interior environments, there is a concern that textiles may cause indoor air pollution. However, textiles can positively affect the interior environment. First, textiles are used in a variety of ways in our interior environments. Textiles are used for upholstery, window treatments, cushions, wall coverings, linens, bed coverings, carpeting, and area rugs. The softness of textiles creates physical comfort within the interior environment (Allen *et al.*, 2004). Second, since textiles are soft, they absorb sound (Ireland, 2007). Third, textiles create a visually and aesthetically pleasing interior, which creates interiors for sensory comfort – touch, sound, visual, and aesthetic (Allen *et al.*, 2004).

On the other hand, some textiles may negatively affect our indoor environment. Many textiles are produced from petroleum products (chemicals) or with the use pesticides. Chemicals and pesticides from the textiles adversely affect indoor air quality (IAQ). IAQ is defined as the condition of the air inside buildings, including the extent of pollution caused by chemicals from materials and appliances, smoking, dust mites, mould spores, radon, and other gases (EPA, 1994; EPA, 1995). Clearly, poor indoor air quality may have negative consequences on the health of the occupants.

An illness that is related to poor indoor air quality is *multiple chemical sensitivity* (MCS). MCS is an illness in which an individual becomes sensitive to various chemicals and irritants at very low concentration levels, and multiple organ systems in the body are affected (EPA, 2003). Exposure to various chemicals found within the environment may trigger symptoms of MCS. The major cause of the illness is poor IAQ.

This chapter will discuss the concerns for indoor environmental pollution (types, sources, and effects), illness related to poor IAQ, problems with textiles and methods of minimizing pollution through textiles, application and examples (selection criteria with a chart and explanation), and future trends (possible changes, demands, and regulations).

8.2 Concerns for indoor environmental pollution

Pollutants that are brought into or that grow within interiors affect the quality of the indoor environment. Pollutants can be classified as microbial contamination, dust mites and animal dander, and volatile organic compounds (VOCs).

8.2.1 Microbial pollution

Microbial pollution comes from mould or mildew – another common term for mould (Air Quality Education, 2007b; EPA, 2007b). Mould is an important part of the environment (EPA, 2007b). Mould lives 'in the soil, on plants, and on dead or decaying matter' (EPA, 2007b). Mould is essential in the outdoor environment; it breaks down leaves, wood and other plant debris. To reproduce, moulds produce tiny spores similar to the way plants produce seeds. These spores are found in the indoor air as well as the outdoor air, and they settle on surfaces in both environments. If spores settle on a damp, wet surface, they will grow and, to survive, will digest the surface. In essence, they need moisture to grow and will ultimately destroy the surface upon which they grow (EPA, 2007b).

In interior environments, mould growth is most commonly found on bathroom tiles, on basement walls, and near leaky water fountains or sinks.

Moisture problems and uncontrolled humidity lead to mould growth and are a greater problem in hot, humid climates (EPA 2007b). Carpeting is one textile that may be affected by mould and mildew, particularly if it is installed in a basement, where moisture is more common. Regardless of the type of fiber (e.g. wool, nylon), if there is moisture, mould can grow (CRI, n/d).

Adverse health effects from mould and mildew are rare. The majority of people have no reaction to mould exposure. If reactions do occur, they will consist of flu-like symptoms and skin rash, which are similar to pollen or animal allergies. However, mould can aggravate asthma, and for individuals with serious immune disease, building-associated mould may cause fungal infections. Generally, symptoms are temporary and may be eliminated by removing or correcting the mould problem (Air Quality Education, 2007b).

8.2.2 Dust mites and animal dander

Dust mites and animal dander become problems to the indoor environment when they become airborne. Many common indoor activities can increase the release of dust, and thus increase exposure to indoor pollution. Such activities that release dust are during vacuuming, dusting, making beds, moving cushions on furniture, or disturbing textiles in various other ways (Science Daily, 2004, MSU-Extension, 2007). Dust mites are of great concern for the asthma sufferer and are the greatest trigger for asthma attacks. Thus, for an asthma sufferer, the best solution is to replace carpet with a hard surface floor and use area rugs that may be removed and cleaned. For other textiles, washable fabrics would be the best solution (MSU-Extension, 2007).

8.2.3 Volatile Organic Compounds (VOCs)

The greatest problem for IAQ occurs from exposure to Volatile Organic Compounds (VOCs). VOCs are either natural or synthetic organic compounds that vaporize at room temperature (Godish, 2001). There are thousand of VOCs; they include formaldehyde, benzene, toluene, methylene chloride, xylene, ethylene glycol, texanol, and 1,3-butadiene (MDH, 2007).

VOCs are found in pesticides, solvents, cleaning agents, building materials, interior finishes and furnishings, adhesives and many other sources (EPA, 1994; Gist, 1999; Thivierge, 1999; Wittenberg, 1996). Among the VOCs, formaldehyde is a chemical that is widely used in industry to manufacture building materials and numerous household products. It can also be found outdoors as a by-product of combustion and some natural processes. Related to textiles, this chemical adds permanent-press qualities to clothing and draperies (EPA, 2003; MDH, 2007).

Most products containing VOCs will emit chemical gases into the air within a short period of time; however, trace amounts of VOCs will continue to be emitted for a long time (MDH, 2007). VOCs enter interiors from various outdoor and indoor sources. Once in the interior, many materials such as carpet, upholstery fabrics, etc. act as sinks for these chemicals gases. After other chemicals have dissipated, these materials and textiles that have absorbed the VOCs will release the chemicals, gases or pollutants back into the air. (Zhao *et al.*, 2002)

VOCs may adversely affect short- and long-term health indoors because there is a greater concentration of VOCs indoors than outdoors (EPA, 2007a). Some VOCs give off an odor that is noticeable in the interior; other VOCs are odorless (Cone and Shusterman, 1999). In particular, formaldehyde – a colorless, pungent-smelling gas – can cause watery eyes, a burning sensation in the eyes and throat, nausea, and breathing difficulties at elevated levels. (MDH, 2007).

Exposure to high concentrations of VOCs may trigger symptoms of MCS. Exposure may occur in one of two ways. First, an individual may be exposed to high doses of a chemical within a short period of time and becomes ill, or second, exposure may occur in two phases. In the first phase, an individual may be exposed to low doses of a chemical over a long period of time, and then later, the individual may be exposed to another chemical or substance that will trigger the symptoms of MCS and, then, will become ill (Gist, 1999; Miller, 1994; Nussbaumer, 2004; Thivierge, 1999).

Exposure to chemicals may take place in a variety of areas. One area of exposure may be in a poorly ventilated building or in communities with high air or water pollution. Another area of exposure may be in unique situations such as during the spraying of pesticides on lawns or fields. Another area of exposure may be in airtight buildings with poor IAQ. (Thivierge, 1999). The latter has become the greatest concern for many of our interior environments.

Poor IAQ also causes a problem known as Sick Building Syndrome (SBS). SBS is defined as a variety of symptoms that affect some of a building's occupants while they are in the building and diminish or go away when they leave the building, yet symptoms cannot be traced to specific pollutants or sources within the building (EPA, 2003). The difference between SBS and MCS is that the individual with MCS will continue to suffer after they leave a building.

The development of MCS may occur from a synergistic reaction between chemicals. This means that one chemical mixing with another chemical creates a reaction with the body that is greater than that which would be expected based on the reaction of each chemical alone (Bower, 2001). Thus, a combination of chemicals from various materials and activities within the built environment can be factors that may trigger symptoms of MCS. However, to date, most studies have been conducted on single chemicals; few studies have been conducted on the health effects of exposure to combined chemicals (MDH, 2007).

8.3 Problems with textiles and methods of minimizing pollution through textiles

Textiles chosen for interiors may greatly impact the IAQ (AIA, 2005). Thus, to minimize indoor air pollution, textiles must be researched prior to selection. Once the appropriate textile has been selected, measures must be taken during installation and continue through maintenance to ensure that IAQ is not compromised. Background should be provided on the characteristics of synthetic and natural fibers related to IAQ, application of finishes, and installation and maintenance of textiles – all relative to IAQ.

Textiles are used for many applications and products in interior environments. Textile fibres may be categorized as synthetic and natural; both have positive and negative aspects relative to the environment (AIA, 2005).

8.3.1 Synthetic fibers

Cellulose-based (semi-synthetic) fibers are derived from the cellulose in cotton and wood, and when they are produced, the physical and chemical properties of cellulose are changed (Allen *et al.*, 2004). These fibers – rayon, acetate, and triacetate – are most frequently used for drapery fabrics because of their draping qualities (Binggeli, 2007). Additionally, they are used in bedspreads, blankets, curtains, sheets, slipcovers, tablecloths, and upholstery. These fibers burn quickly and easily (Jackman *et al.*, 2003). When some cellulosic fibers burn, they give off carbon dioxide and carbon monoxide. Carbon dioxide can cause suffocation, but carbon monoxide will poison the blood stream (Health & Safety, 2006). For this reason, a flame retardant finish is applied to such fibers for residential use (Nielsen, 2007).

Synthetic man-made fibers are produced from petroleum products. Synthetic fibers – nylon, polyester, acrylic, modacrylic, olefin, saran, vinyl, and vinyon – are the most widely used for interior finishes and furnishing such as carpet, upholstery, and draperies. Synthetic fibers are strong, and manufacturers are able to shape them at high temperatures. Most will melt when in contact with heat such as a hot iron or a flame (Binggeli, 2007;

Encyclopædia Britannica, 2007a,b). When the flame is removed, each fiber reacts differently to flame. Nylon and polyester will self-extinguish; olefin will continue to burn and slowly melt; acrylic and vinyl melt and burn slowly (Neilsen, 2007); however, the greatest concern is the toxic fumes (carbon monoxide and other toxic gases) given off (Health & Safety, 2006). Again, flame resistant finishes are used to retard a fire.

When polyester melts, it gives off toxic fumes (Jackman *et al.*, 2003). Thus, to prevent this from occurring, a flame retardant finish is applied (Nielsen, 2007). In the 1970s, synthetic fibers (polyvinyl alcohol and polyvinyl chloride) were produced with a flame retardant that remained even through laundering or dry cleaning. These fibers produced an extremely low toxic fume when burned; this was particularly important to minimize pollution in interiors and in a fire. These fibers have been used for home furnishings (carpeting, mattresses, and blankets) as well as apparel and industrial fabrics. Fiber characteristics included a soft textural feel, abrasion resistance, and moderate moisture regain (Koshiro, 1974).

Since that time new materials have been produced. One example is an aromatic polymer-based manufactured fiber called polybenzimidazole (PBI). PBI neither melts nor burns and, when charred, remains intact and supple. With these properties, toxic fumes have been eliminated and the material is safe to use in aircraft, hospitals, and submarines for upholstery, window treatment fabrics, and carpets (Binggeli, 2007).

Eco-Intelligent™ Polyester was introduced in 2001. This was the first polyester to be produced and dyed with all environmentally safe ingredients. *Eco-Intelligent™ Polyester* was developed in partnership with McDonough Braungart Design Chemistry (MBDC) and its German sister company, Environmental Protection Encouragement Agency (EPEA). It was also designed to be safely recycled into new fabric at the end of its life without the hazardous by-products of traditional polyester recycling. It is not only a revolutionary fabric but also a healthy alternative for the textile trade (McDonough and Braungart, 2002b).

Unfortunately, not all synthetic fibers have this quality. Polyolefin properties (lightweight, elastic, resist stains and crushing) have made it a good fiber for carpet, upholstery, and draperies. However, polyolefin shrinks and melts when heated (Binggeli, 2007).

8.3.2 Natural fibers and organically grown fibers

Natural fibers originate from either cellulose (plants) or protein (animals). They include cotton, flax, linen, hair, fur, silk, and wool (Encyclopædia Britannica, 2007b). It is often assumed that natural fibers would be safer than synthetic fibers. However, some natural materials have a negative affect on the outdoor environment as well as indoor air pollution (Haberle, n/d; Mate, *et al.*, 2004). Though cotton is a natural fiber, ten percent of the worlds' annual pesticide consumption is used in its production (Haberle, n/d; Mate *et al.*, 2004).

Natural cotton is also highly treated. Industrial chemicals such as formaldehyde, chromium, and chlorine are used for bleaching, dyeing, and finishing (Bonda and Sosnowchik, 2007). Such applications to the surface further create indoor environment pollution. To minimize pollution, the better alternative is organically grown fibers with minimal treatment.

Organically grown cotton must meet specific certified standards. These standards require that no synthetic or chemical pesticides or fertilizers can be used on the land or in the growing process. Once harvested, the cotton must be washed only with a mild soap and no bleach applied. Though the natural color of cotton textiles is preferred, some dyes may be used (Binggeli, 2007). Clearly, such textiles will minimize indoor air pollution.

Though wool is a natural fiber, there are some concerns for IAQ. The first concern is controversial. According to the Environmental Building News (EBN) (1994), wool fibers have a greater potential for trapping VOCs (especially formaldehyde and nitrogen oxides) than synthetics. However, in other studies, it was determined that wool carpet may even purify the air by absorbing these indoor pollutants (formaldehyde, nitrogen dioxide and sulfur dioxide) (Fibres, 2002). Research needs to be conducted to determine which is correct or if it is situational.

The second concern is mothproofing of wool fibers. According to EBN (1994), mothproofing is environmentally safe – regarding both outdoor and indoor pollution. However, the mothproofing is a chemical additive, which may affect IAQ. The best solution to minimize indoor pollution is to use untreated wool; however, untreated carpet or other untreated textiles may be hard to find (EBN, 1994).

The third concern is that dust particles can affect allergy sufferers. However, wool fibers' microscopic scales will capture and hold dust particles until vacuumed. This results in less dust, which is helpful for allergy sufferers (Fibres, 2002).

8.3.3 Biodegradable, recycled, and/or recyclable fibers

Natural plant and wool fibers are biodegradable as well as being a renewable resource (Fibres, 2002). Recycled materials may be from one of two sources: (i) pre-consumer or (post-industrial) waste and (ii) post-consumer waste. Using pre-consumer waste to produce other products reduces waste placed in landfills (Nielsen, 2007). Post-industrial recycled yarns from the textile industry (wool, cotton, etc.) can be used to produce a variety of interior textiles such as mattresses, coarse yarn, home furnishings, and other products. Post-consumer materials such as PET bottles can be turned into yarns (Nielsen, 2007); this is one way to preserve valuable resources such as petroleum (Capek and Grose, 2005).

8.3.4 Dyes and treatments

Chemical finishes may also be used on textiles (carpeting, upholstery, drapery fabrics) for antistatic, antimicrobial, stain resistant, wrinkle resistant, or flame resistant purposes (Bingelli, 2007; EBN, 1994). However, these processes not only use chemicals that affect air quality but also use vast amounts of water and energy. Though industry insists that these treatments are safe, many consumers are concerned about air quality related to such chemical treatments (EBN, 1994). Many standard finishes, such as an antibacterial or antiseptic finishes, are chemicals and some may be formaldehyde-based such as the permanent press finish (Nielsen, 2007). Bleaches, dyes, fabric protectors, and fire retardants may emit VOCs into the interior, and thus contribute to indoor air pollution (AIA, 2005).

Wool carpet fibers are naturally stain-resistant, repelling liquids and most stains (Fibres, 2002; Nielsen, 2007). Generally, spills are easily blotted up and dirt is removed efficiently with regular vacuuming. However, wool fiber with no chemical stain resistant treatments will stain more readily than many synthetic fibers with stain resistant finishes. Wool carpet fibers respond well to professional wet-cleaning (Fibres, 2002). Likewise, polyester carpet fibers are inherently stain resistant and also respond well to professional carpet cleaning (Fibres, 2002). Nylon fibers with stain resistant finishes resist damage from oil and chemicals and may be dry-cleaned or wet-cleaned.

Nylon fibers are also treated with a static-resistant finish. However, polyester carpet fibers are less likely to generate static electricity and, thus, do not requirement a static resistant finish. Wool fibers also do not require a static resistant finish as wool retains moisture with the fabric that prevents a build-up of static electricity (Fibres, 2002; Wool Fibre, 2007).

Fortunately, new safe finishes are now added or applied to fibers or textiles. Biodegradable finishes are replacing the potentially hazardous chemical finishes (Bingelli, 2007). Natural dyes may seem a better alternative to synthetic; however, very toxic metallic compounds are needed to make some natural dyes more stable (Bonda and Sosnowchik, 2007). However, non-chemical finishes are available. *FoxFibre®* and *Colorganics* are organically grown and naturally colored cotton textiles. These textiles are created to be truly environmentally safe – no bleaches, dyes, or chemical finishes are used (AIA, 2005). More information can be found about these materials on the website: http://www.vreseis.com.

8.3.5 Flame resistance

Wool carpet fibers will not ignite, and, if exposed to flame, will self-extinguish quickly and leave the carpet with a small amount of ash that is easily brushed away. Conversely, a single cinder or spark will melt many synthetic carpet fibers and leave a burnt patch (Fibres, 2002). Thus, a flame resistant finish is commonly applied to commercial grade synthetic carpet. Flame resistant finishes make products less likely to catch fire and, therefore, are important in saving lives (EBN, 2004). The retardants most frequently added to various consumer products are brominated flame retardants (BFR). Recently, the use of BFR has become a concern for the environment as well as the human health (Birnbaum and Staskal, 2004; EBN, 2004). With concerns for BFR, some manufacturers have eliminated these from their products. The office furniture manufacturer, Herman Miller is attempting to eliminate the chemical in their new products and is redesigning existing products such as the Aeron and Mirra chairs. The carpet tile and fabric manufacturer, Interface is working to eliminate BFR at its Canadian carpet tile plant and LaGrange, Georgia plant (EBN, 2004).

Since many fire retardant finishes are synthetic chemicals, avoiding combustible materials is advised whenever possible to improve indoor air quality (Bonda and Sosnowchik, 2007). Thus, aramid is an alternative fiber to those with chemical treatments. Aramid is a manufactured fiber with inherent high temperature and flame resistance (Yeager and Teter-Justice, 2000). It can be laminated to the back of upholstery fabrics. It meets the high standards of California Technical Bulletin 133 (CAL 133) and the lamination of aramid fabrics protects upholstery foam from fire (Bingelli, 2007). Because of their flame resistance, aramid fibers are often used in aircraft upholstery and carpet (Yeager and Teter-Justice, 2000).

8.3.6 Maintenance

With improved energy efficiency in air-tight buildings, indoor air is frequently recirculated. This results in the build-up of indoor air pollutants such as tobacco smoke, VOCs, biological contaminants, and allergens from dust mites. Textiles (carpet, draperies, upholstery, etc.) absorb and trap pollutants. As pollutants build up, they are emitted into the air. Therefore, proper and scheduled maintenance play a crucial role in controlling indoor air pollutants (IAQ, 2006). Once products and/or materials are installed, a maintenance schedule should be developed. Additionally, cleaning products should not contain either VOC or formaldehyde-related chemicals to minimize indoor air pollutants and maintain good IAQ. Many cleaning products are certified by GreenGuard and can be found at http://www.greenguard.org.

8.3.7 Dry cleaning

Some textiles require dry cleaning. However, solvents used in dry cleaning will adversely affect IAQ. According to Air Quality Education (2007a), trichloroethylene (perchloroethylene, PCE) is a common solvent used in the dry cleaning process. The solvent residue from dry cleaning generally remains on the fibers and will emit odors into the interior. Another service that many dry cleaners perform is adding mothproofing chemical to wool items. This was discussed earlier in this chapter. Thus, dry cleaning chemicals and moth proofing can affect IAQ and, particularly, individuals with MCS.

8.3.8 Clean environment

A clean environment is important to an individual whose chemical sensitivities include allergies to dust, mould, and mildew. To limit the dust, dusting twice a day to keep dust particles at a minimum is recommended. Dehumidifiers are also recommended to eliminate moisture and the development of mould and mildew. The interior environment must also be free of VOC toxins. This is a greater challenge with the abundance of chemicals used in construction and applied to materials.

8.4 Applications/examples

To select the appropriate textiles for interior environments, it is important to examine the expectations that will provide for a healthy environment.

8.4.1 Expectations and/or concerns in selecting appropriate textiles

The following is a list of expectations and/or concerns in selection the appropriate textiles that will minimize indoor air pollution:

- Select untreated natural and organically grown fibers whenever possible.
- Select textiles that are dyed with organic or natural pigments that are free of toxins, carcinogens, and heavy metals.
- Select textiles that are solution dyed versus conventional dyeing.
- Select textiles that are inherently fire retardant and stain resistant rather than needing chemical treatments.
- Select textiles with low or no-VOC content.
- Select textiles with high-recycled content.
- Select biodegradable textiles.

- Maintain low humidity to lessen opportunity for mould and mildew to form.
- Maintain proper ventilation.
- Use Manufacturers' Material Safety Data Sheets (MSDS).
- Become more aware of toxic and hazardous material content (AIA, 2005; Bonda and Sosnowchik, 2007; Wright, 2006).

8.4.2 Selection criteria

The selection criteria chart (Table 8.1) provides a format that may be used to appropriately select the textile and, particularly, select those that will minimize indoor air pollution. The first column lists the criteria. The second column is the expectation. A question to ask is 'what is the optimum expectation for this criterion?' The last column is filled from the specifications of the textile after it is located. In writing these specifications, it is

Table 8.1 Selection criteria form related to indoor air quality

Textile type: Upholstery *[Insert the textile to be selected, e.g. Carpeting, upholstery, etc]*

Criterion	Expectation	Material specification
Fiber type	Cotton [as an example]	(Specification of textile selected listed in this column)
Production process	Organically grown	
Treatments		
Stain resistant	Inherent	
Static resistant	Non-toxic (Low or no VOC)	
Anti-microbial	Non-toxic (Low or no VOC)	
Mold/Mildew resistant	Non-toxic (Low or no VOC)	
Fire resistant	Inherent or non-toxic (Low or no VOC)	
Toxic chemicals		
VOCs emissions	Non-toxic (Low or no VOC)	
Formaldehyde	None	
Recycled content	If possible	
Maintenance	Non-toxic (Low or no VOC) Cleaning products Develop a maintenance schedule	
Reclamation program	Require a recycled program	
Biodegradable	Must be biodegradable	

Table 8.2 Textile types with general recommendations and web addresses

Type of textile	Recommendations	Web address
Floor coverings		
Carpet and area rugs	• Low emitting carpet, carpet pad, & adhesives • IAQ & Green labeling • Appropriate installation process	http://www.carpet-rug.com/index.cfm
Upholstery, textile panels, textile wall coverings		
Upholstery, panels, textiles wall covering	• 100% organically-grown • Pure wool	http://www.carnegiefabrics.com/ http://www.dtex.com
Mattresses and bedding	• 100% rubber • 100% organically grown • Pure wool	http://www.nontoxic.com http://furnature.com http://satara-inc.com http://www.suitesleep.com
Various materials and products		
General	• Select supporting materials and products that promote good IAQ	http://www.greenguard.org http://www.greenseal.org http://www.usgbc.org

important to determine if the expectations (listed in the second column) are met. If so, the product will minimize indoor air pollution. If not, there may be two reasons: (i) no textile meets all criteria, and (ii) further investigation into other textiles is needed. Also in considering the criteria, it is important to consider the production process, type of finish or treatment, and avoidance of toxic chemicals in all aspects of a textiles use.

Textiles with recommendations for applications are included in chart form (Table 8.2). The first column includes the type of textile such as carpet, upholstery, textile panel, and textile wall covering. The second column is a list of recommendations, and the third column provides web addresses for more information regarding the various types of textiles.

The following is a discussion of various textile applications and will provide information regarding minimizing indoor air pollution as well as examples of materials and products that are available that meet specific needs.

8.4.3 Soft floor covering: Carpet

Carpet is one of the most common floor coverings used in residential and commercial interiors. Carpet is made of synthetic fibers (nylon, polyester, or polypropylene) or from natural fibres (wool or cotton). Synthetic nylon is the most common fiber used for carpet; a polyolefin such as polypropylene is an alternative to nylon because of its affordability and wool-like feel. Polyester is installed in low-traffic residential settings, as it stains and crushes more easily. Wool carpet is not as frequently installed because it is a more expensive fiber. However, it lasts longer making it more affordable in the long term (Nielsen, 2007). Also wool is non-toxic and renewable (AIA, 2005). Conversely, if processing includes the use of pesticides, or dyes are applied, wool loses it safe, environmental qualities. This is also true if the carpet is constructed with toxic adhesives, or stain repellents containing formaldehyde are applied.

Because of its petroleum base, synthetic carpet has been considered a major source of triggers for individuals with MCS. Conversely, Igielska *et al.* (2002) found that the emissions of VOCs were very low; and yet in the Igielska *et al.* (2003) follow-up study, they found that the VOC emissions from new floor coverings increased as the temperature increased. Over time, VOC contamination decreased into the interior environment. However, over time, as discussed earlier, the sink effect may take place and VOC may still be absorbed from other sources and released into the indoor air.

In many cases, the carpet itself is not the problem. Sometimes it is the finish applied to the carpet (e.g. chemicals used for stain resistance, flame resistance). Sometimes it is the carpet pad and/or adhesive that contains chemicals. Sometimes it is the process of installation that affects IAQ. Other times it is poor maintenance – dust, dirt, mould, and/or mildew will affect IAQ and trigger symptoms of MCS or asthmatic attacks. Clearly, chemicals, dust fibers, moulds, bacteria, and metals from carpet are released into the indoor air. Additionally, old synthetic carpeting, which becomes brittle with age, gives off synthetic house dust (Air Quality Education 2007a). Thus, to minimize indoor pollution, the appropriate carpet must be selected. To do so, the carpet manufacturer's representative should be contacted for information on low emitting carpet as well as carpet pad and adhesives. Information regarding carpet materials, IAQ, and 'Green Label' testing may also be found at http://www.carpet-rug.com.

The process and method of installation may also prevent the emission of chemicals. Prior to installation, it is important to unroll and air out the new carpet in a clean, well-ventilated area. When installation begins, the individual with MCS, asthma, or a similar illness should not enter the space during or immediately after the installation. This will allow time for toxins to be released and exhausted from the space. During the installation, doors and windows should be opened to increase fresh air. This will reduce the chance for installers and other occupants to be exposed to chemicals released from newly installed carpet. Also, during and after installation, window fans and room air conditioners should be running to increase the amount of fresh air and also to exhaust toxic fumes to the outdoors. Ventilation systems should be in proper working order and in operation during installation as well as for 48 to 72 hours after installation (EPA, 1994, p. 21). Lastly, once the carpet is installed, a maintenance schedule and list of recommend chemical-free cleaning products should be prepared.

Disposal of old carpet creates billions of pound of waste yearly. Thus, selecting a biodegradable fiber is a wise decision. Natural fibers are biodegradable. For example, carpets of plant fibers as well as the jute backings are completely biodegradable as well as a renewable resource. Wool carpet fibers are also biodegradable or may be recycled into another material (Fibres, 2002; Organic and Healthy, 2007). Wool fibers that are processed without chemicals will minimize indoor air pollution (Organic and Healthy, 2007).

Also, in the foreseeable future, biodegradable electropsun cellulose mats will be used to absorb fertilizers, pesticides, and other materials. Then, later at a desired time and location, the materials would be released to a safe receiver (Lang, 2003). This may reduce the release of contaminants into the interior.

Colin Campbell and Sons from Vancouver, Canada are manufacturers of a non-toxic wool carpet – *Nature's Carpet* – that has been processed without the use of pesticides or dyes. Its backing is also a plant-based latex jute (AIA, 2005).

Recycled carpet fibers can be made from polyester fibers that have been recovered from plastic bottles. Creating fibers from recycled bottles begins by collecting bottles and packing them into bales to ship to a plastics recycling plant. At the plant, bottles are sliced and ground into chips. Then, the chips are washed and blown free of paper labels. Next, the chips are shipped to a factory where they are melted and forced through spinnerets that extrude thin fibers. Then, fibers are spun and tufted as if they were virgin polyester (Fibres, 2002).

Image Carpets from Armuchee, Georgia manufacture ENVIRO-TECH carpet, also from *fibre* derived from post-consumer soft drink bottles. It has the lowest VOC emissions of PET carpet on the market (AIA, 2005).

To encourage recycling of carpet, carpet manufacturers such as Invista have developed a reclamation program (INVISTA™ Carpet Reclamation Program). This effort minimizes the total environmental impact. Though this reclamation program does not directly affect the pollution of

Table 8.3 Sources for carpet

Product	Origin of fiber	Notes	Manufacturer and/or supplier
Carpet: Needle-punch	Pre-consumer textile waste	100% recycled content, biodegradable, non-toxic	Reliance Carpet Cushion
ENVIRO-TECH	Recycled PET	Post-consumer soft drink bottles	Image Carpets
Carpet	PET	Lowest VOC emissions of PET carpet	Armuchee, GA
Carpet	Synthetic fibers reclaimed	Reclamation program	Antron Reclamation Program™
Carpet	Non-toxic wool	Renewable Processed without pesticides or dyes. Plant-based latex jute backing	Colin Campbell and Sons Vancouver, Canada

the interior, it is important for the low impact on the environment (Antron Carpet Fibre, 2007)

Reliance Carpet Cushion Division are manufacturers of needle punched carpet cushions made from pre-consumer textile waste fibres and recycled carpet. This carpet is labeled as 100% recycled content, biodegradable, and non-toxic. Their cushions are also mould- and mildew-proof, and non-allergenic (AIA, 2005). Table 8.3 provides a recap of information on carpeting.

8.4.4 Soft floor covering: Area rugs

Area rugs are popular and can be used over hard or resilient surfaces. Where carpet will receive heavy traffic, a washable rug is the best choice. Where there is less traffic, the choice of carpet should be researched in a similar fashion to broadloom carpet. Fibers used for broadloom carpet may be used for area rugs; however, rayon, cotton, wool, and acrylic are the most common fibers used (Nielsen, 2007).

Allure Rug Studio designs and produces hand-tufted wool area rugs. To color the yarn, only all-natural dyes are used. Cotton is used for both primary and secondary backings, but the backing is adhered with a latex glue (AIA, 2005). More information may be found at http://allurerugstudio.com.

Table 8.4 Sources for area rugs

Product	Origin of fiber	Notes	Manufacturer and/or supplier
Hand-tufted area rugs	Wool Cotton	Only natural dyes used Cotton for primary and secondary backing Latex glue adheres backing	Allure Rug Studio designs http://allurerugstudio.com
Natural fiber and textile floor covering	Sisal, hemp, jute, coir, bamboo, seagrass, and wool		Merida Meridian, Inc. Boston, MA http://www.meridameridian.com
Area rugs and broadloom carpet	100% wool	Moth-proof New-Zealand wool Cotton-backing coated with a water-based limestone latex	Savnik & Company Oakland, CA

Merida Meridian, Inc. from Boston, Massachusetts, create natural fiber and textile floor coverings that are made from sisal, hemp, jute, coir, bamboo, seagrass as well as wool (AIA, 2005). More information may be found at http://www.meridameridian.com.

Savnik & Company from Oakland, California, creates custom 100 percent wool area rugs as well as wool and cotton wall-to-wall carpet. The rugs are woven using a moth-proof New Zealand wool and a cotton backing that is coated with a water-based limestone latex (AIA, 2005). Table 8.4 provides a recap of information on area rugs.

8.4.5 Upholstery, textile panels, textile wall covering and draperies

Fibers used for upholstery include wool, polyester, olefin, nylon, cotton and leather, as well as blends of fibers. Fibers also may be derived from recycled materials. Arc-com *Eco-Tex* textiles are recycled textiles that may also be recyclable at the end of their useful life. The textiles are woven using

100 percent post-industrial or post-consumer recycled polyester. If a backing is used, it is also from recycled materials. *Eco-Tex* textiles may be used for both upholstery and panel fabric. Also, *Eco-Tex* textiles are produced in a sustainable manufacturing process in which waste, emissions, energy consumption, water usage, and toxic substances are reduced. Arc-com is a Green Seal® (2005) environmental partner (AIA, 2005). More information may be found on the website: http://www.arc-com.com.

Designtex and Carnegie fabrics have textiles in their line that are totally organic and environmentally safe. Such products have not only been produced and manufactured using environmentally safe practices but are also safe for indoor air quality (Carnegie, n/d; Designtex, 2005). Designtex supply commercial fabrics for a variety of uses: upholstery, panels, wall covering, and drapery. Some of their fabrics have been produced to be ecologically intelligent. For example, *Eco-Intelligent Polyester* (discussed earlier) is a synthetic textile produced without the use of antimony-based catalysts; rather it is produced using an environmentally-safe titanium catalyst. Additionally, fewer dye chemicals as well as less water and energy are used in the manufacturing process. This process allows the textiles to safely return to a natural state. The entire process of production to the end of its useful life has been addressed. In partnership with Steelcase, Designtex has also launched the *Environmental Impact Collection* (100 percent recycled polyester). 'These fabrics are designed to be used without chemical backings and are manufactured with processes that conserve the earth's resources' (AIA, 2005). They have been created to be recycled as well as to work with the latest Steelcase seating and panel fabrics. These textiles are not only resource efficient, recyclable, biodegradable and durable but also have low toxicity (AIA, 2005; Designtex, 2005). More information may be found at http://www.dtex.com.

Climatex® *Lifecycle*™ textiles are made from natural fibers (wool and ramie) that wick away moisture; they are breathable, and naturally stain-resistant. Climatex® *LifeguardFR*™ textile is not only an environmentally optimized material but also meets the strictest flame retardant standards throughout the world. They are made from wool and a regenerated cellulose fibre *LenzingFR*™–harvested from renewable beech wood (Designtex, 2005; Carnegie, n/d).

Xorel® textiles are made from polyethylene and are free of chlorine and plasticizers. Xorel® is inherently stain-resistant and flame retardant, with no chemical treatments applied. This textile is designed for use as wall treatment, fabric panels, and upholstery. As a low- or non-toxic textile, it minimizes pollution on the interior (Carnegie, n/d).

Terratex brand recycled polyester is made of 100 percent recycled or renewable fiber that is made from post-consumer and post-industrial recycled yarns. This type of product reduces the need for virgin polyester, which saves resources and energy (AIA, 2005; Designtex, 2005). More information may be found at http://www.dtex.com, http://www.carnegiefabrics.com, http:// www.guilfordofmaine.com, and http://www.terratex.com.

Cargill Dow, LLC produces *Ingeo* fiber that is made from corn – a product that is a 100 percent annually renewable resource. It is also naturally flame retardant and energy efficient, recyclable, renewable, as well as low in toxicity to minimize indoor air pollution (AIA, 2005). More information may be found at http://www.cargilldow.com.

Deepa Textiles also emphasize sustainability (recycled content and resource efficiency) as well as creating a fiber that is low in toxicity (AIA, 2005). An example of one of Deepa's textiles is *Glassworks* vertical-surfacing collection that is designed for contract and health-care installations. This product is created from a recycled *Terratex* polyester (Glassworks, 2003). More information may be found at http://www.designjournalmag.com/product/Glassworks.htm and http://www.deepa.com.

Hemp Textiles International Corporation (HTI) develops, manufactures, and sells hemp textile products; their exclusive brand is a luxury fiber (*Hemptex Cantive*). All products in the HTI line are produced without chemicals or fertilizers. Because hemp is a plant fiber, it is biodegradable and recyclable. More information may be found at http://www.hemptex.com.

Maharam provides products that reduce environmental impact. Some textiles have natural fiber content such as cotton and wool; some have recycled content (post-industrial and post-consumer recycled polyester); some are biodegradable; many are free of heavy metal dyes. However, the fibers are not entirely free of toxins with the use of natural cotton (not organically grown). They are classified as having an 'environmentally improved' manufacturing process. More information may be found at http://www.marharam.com. Table 8.5 provides a recap of information on upholstery, fabric panels, wall covering, and drapery textiles.

8.4.6 Finishes

Crypton® is a textile whose weaving process is coupled with stain, water, and bacteria resistant barriers. *Crypton®* fabrics contain neither VOCs nor formaldehyde, and no perfluorooctanoic acids (PFOAs) are used during the manufacturing process. It also passes the CAL 133, UFAC Class 1 and other fire code tests. *Crypton® Green* (2007), which is available through Momentum's collection, is made from 100 percent recycled polyester. This high-performance fabric is low-emitting and is not only recycled but is recyclable (Crypton®, 2007). More information may be found at http://www.cryptonfabric.com.

Table 8.5 Sources for upholstery

Product	Origin of fiber	Notes	Manufacturer and/or supplier
Upholstery Panel fabric	100% post-consumer or post-industrial recycled polyester	If backing used, must be from recycled materials	Arc-com Eco-Tex http://www.arc-com.com
Eco-Intelligent Polyester	Polyester		MDBC & Designtex http://www.dtex.com
Environmental Impact Collection	100% recycled polyester		Designtex http://www.dtex.com
Climatex® Lifecycle™	Wool and ramie	Wicks away moisture, breathable, and naturally stain resistant	http://www.dtex.com
Climatex®	Cellulose fiber	Environmentally optimized material	http://www.dtex.com
LifeguardFR™	LenzingFR™	Meets strictest flame retardancy standards Wool and renewable beechwood	
Xorel® textiles	Polyethylene Free of chlorine and plasticizers	Inherently stain resistant and flame retardant with no chemical treatments applied	http://www.carnegiefabrics.com

Terratex	Recycled polyester	100% recycled or renewable fiber from post-consumer and post-industrial recycled yarns Product reduces need for virgin polyester and saves resources and energy	http://www.dtex.com http://www.carnegiefabrics.com http://www.guilfordofmaine.com http://www.terratex.com
Ingeo fiber	Corn	100% annually renewable; natural flame retardant and energy efficient, recyclable, renewable; low in toxicity	Cargill Dow, LLC http://www.cargilldow.com
Deepa Textiles	Recycled content	Recycled content and resource efficiency; low in toxicity	http://www.deepa.com
Hemptex Cantive	Hemp	Plant fiber produced without chemicals or fertilizers, biodegradable, recyclable	Hemp Textiles International Corp. (HTI), http://www.hemptex.com
Maharam	Natural, recycled, and/or biodegradable	Various types: natural fiber content; recycled content; biodegradable; many free from heavy metal dyes Natural cotton is not entirely free of toxins (not organically grown)	http://www.marharam.com

Table 8.6 Finishes and backings for other products

Product	Origin of fiber	Notes	Manufacturer and/or supplier
FoxFibre®	Organically grown and naturally colored cotton	No bleaches, dyes, or chemical finishes	http://www.vreseis.com
Colorganics	Organically grown and naturally colored cotton	No bleaches, dyes, or chemical finishes	http://www.vreseis.com
Aramid fabric	Backing for upholstery fabric	Alternative to chemical treatment; meets California Technical Bulletin 133 (CAL 133); protects upholstery foam from fire	
Crypton®	100% recycled polyester coupled with stain, water, and bacteria resistant barriers	No VOCs, no formaldehyde, and no perfluorooctanoic acids (PFOAs) used during manufacturing process Passes Class 1, CAL 133	http://www.cryptonfabric.com

Products discussed earlier in this chapter include FoxFibre® and Colorganics, which contain neither bleaches, dyes, nor chemical finishes. More information may be found at http://www.vreseis.com. Also previously discussed was information on upholstery backings using aramid fabrics. Table 8.6 provides a recap of information on finishes and backings.

8.4.7 Mattresses and bedding

Sources for organically grown textiles such as cotton and wool are available for mattresses and bedding (Nivana *Safe Haven*, n/d). Though the products are organic, they are not fire resistant. Thus, their safety may need further investigation. More information may be found at http://www.nontoxic.com.

There are several other firms that have environmental safe mattresses, bedding, and upholstered furniture. Furnature produces upholstered organic furniture as well as bedding. Furnature's mattresses and bedding are made from fibers (100 percent organic cotton or chemical-free wool) that contain no chemicals, dyes, polymers, or toxins. Therefore, these products will minimize indoor pollution (Furnature, 2003). More information may be found at http://furnature.com.

Satara products are designed using 100 percent organic cotton, 100 percent natural latex, and/or pure-grown wool. Mattresses are hypoallergenic and contain no dyes or chemicals. Products for home as well as babies are included in their line (Satara, n/d). More information can be found at http://satara-inc.com.

Suite Sleep (2007) represents several lines of organic products for adults and infants. Products include mattresses, bedding, pillows, towels and other linens. Particularly, *Green Sleep* is designed using 100 percent natural rubber latex, and the cover is made of organic cotton, pure wool, and raw silk. More information can be found at http://www.suitesleep.com. Table 8.7 provides a recap of information on finishes and backings.

The application and available product examples will provide a basis for selection of appropriate textiles – textiles that will minimize indoor air pollution and promote good IAQ.

Table 8.7 Mattresses and bedding

Product	Origin of fiber	Notes	Manufacturer and/or supplier
Nontoxic textile	Cotton; wool	Organically grown textiles	http://www.nontoxic.com
Furnature	100% organic cotton, chemical-free wool	Upholstered organic furniture and bedding	http://furnature.com
Home-environment	100% organic cotton, 100% natural latex, and/or pure-grown wool	Hypoallergenic and contain no dyes or chemicals	http://home-environment.com
Green Sleep	100% natural rubber latex; organic cotton, pure wool, and raw silk	Suite Sleep has several lines of organic products	http://www.suitesleep.com

8.5 Future trends

8.5.1 Changes that may occur in the future

Changes that may occur in the future relate to chemicals, testing, and research. Chemicals that create indoor air pollution are often used for textiles in the dyeing process and the application of other finishes such as stain resistance. To minimize indoor air pollution, changes need to take place. For the dyeing process, better natural dyes need to be developed and harmful chemicals should be eliminated from this process. For other finishes, e.g. stain resistant finish inherent or naturally resistant, need to be found and harmful chemicals need to be eliminated from the process. With such changes, textiles will minimize indoor air pollution.

At present, testing for VOC emissions requires sophisticated equipment. For example, scientists at the Georgia Tech Research Institute (GRIA) have developed a room-sized environmental test chamber. This chamber measures ultra-trace concentrations of airborne chemicals being emitted from products. Using the testing chamber, manufacturers are able to choose textiles that have minimal emissions (Toon, 2006). This equipment is important for manufacturers of a variety of materials; however, the ability for a consumer to monitor their own indoor air could positively affect the health of the occupants. In the near future, effective testing instruments may be available to the consumer.

Collaborative international research may aid positive changes. For example, Canesis Nework Ltd from Christchurch, New Zealand is collaborating in a study with the St. Louis School of Public Health and the US government to develop a better understanding of the complex interplay of interior furnishings and particulate air pollutants. In another study, researchers from Canesis and Otago and Victoria University are investigating how allergens and bacteria accumulate in domestic environments (Rewi, n/d). In time and through research, there will be more effective ways to further minimize indoor air pollutants.

8.5.2 Greater demand for non-polluting textiles

Designers have heightened awareness of indoor air pollution, and many have been selecting materials that promote good IAQ. Consumers have also become more aware of the dangers of indoor air pollution, and some are asking for materials that are environmentally safe for the interior as well as to the outdoor environment. To the advantage of both designers and consumers, great strides have taken place to produce materials that minimize indoor air pollution. The number of manufacturers and suppliers of environmentally friendly materials has grown, and a broader and more competitive market for green textiles is taking place.

According to McDonough and Braungart (2002b), many textile makers, fabric distributors, and furniture manufacturers are coming together and exploring the design of a 'take-back program' for textile recycling. For example, the automotive industry has seen the economic benefit of reusing valuable materials and is already moving toward implementing 'take-back programs'. In Europe, the law requires the reclamation of automotive materials. In future, the ability to produce polyester from renewable resources and transform it into a fully biodegradable material is expected. Producing only environmentally friendly materials devoid of toxins could be part of the Next Industrial Revolution.

8.5.3 Regulations

In the past, the textile industry has been considered one of the world's worst polluters. With the variety of fibers and applications – some made from chemicals, others processed with chemicals, and still others with chemical finishes as well as emissions – the amount of pollution is not surprising. However, with increased awareness and legislation, more pressure is being placed on local producers as they look at export opportunities to countries with stringent requirements (Jacka, 2001).

Due to public awareness, there has been an increased demand for government regulation (DustFree®, 2006). In 1994, the United States Department of Labor, Occupational Safety and Health Administration (OSHA) proposed the adoption of standards to address indoor air quality within the work environment. Standards proposed would require employers to implement controls on specific contaminants and their sources – tobacco smoke, microbial contamination, maintenance and cleaning chemicals, pesticides, etc. However, in 2001, this Indoor Air Quality proposal was withdrawn and this terminated the rulemaking proceedings. The proposal was withdrawal because many states and local governments as well as private employers took action to stop smoking in public and work areas (OSHA, 1994; OSHA 2001). Many municipal governments have enacted regulations requiring building owners and operators of restaurants, bars, etc., to control smoking within these buildings (DustFree®, 2006).

Presently, at the national level, OSHA only requires that employers 'furnish to each of [their] employees employment and a place of employment which are free from recognized hazards that are causing or are likely to cause death or serious physical harm to [their] employees'. However, OSHA regulations do required responsible parties to document information on potentially hazardous products. This information is provided within Material Safety Data Sheets (MSDS). And yet, since MSDS are not available for every product or material, the information is limited. Clearly, regulations for IAQ need to be proposed and upheld through OSHA.

According to the South African Bureau of Standards (SABS), a new approach has been taken to create regulatory mechanisms that deal with environmental issues. Rather than using governmental regulations, the SABS has challenged the textile industry to examine their impact on the environment as well as establish objectives and targets, and then to develop processes for the prevention of pollution that would include employees' health and wellbeing (Jacka, 2004). In South Africa, the Council for Scientific and Industrial Research is a leading scientific and technology research organization (CSIR, 2007). The CSIR Textile Technology has developed an eco-labeling scheme called, Oeko-Tex. To receive the label, the textile (clothing, bedding, etc.) must meet stringent standards for a limited level of formaldehyde (Jacka, 2004).

Interestingly, there are labels for biodegradable textiles that are designed using organic dyes and fiber. The question is, are people willing to pay the higher cost for eco-textiles? In other words, people talk about environmentally safe products, but when it comes to cost, many are not willing to pay the extra expense (Rewi, n/d). Hopefully, in time, as more eco-textiles are available, the cost will be reduced.

8.6 Sources of further information and advice

Many resources are available to help designer's select appropriate materials for safe interior environments. These may be found in books, trade magazines, research and interest groups, professional groups, and websites.

8.6.1 Books

McDonough and Braungart (2002a), authors of *Cradle to Cradle: Remaking the Way We Make Things,* suggest eliminating the present concept of the buying, using, and disposing of products. Instead, products should be designed to provide nourishment for something new at the end of its life. Products can be conceived as 'biological nutrients' that will re-enter the water or soil without depositing toxics or synthetic materials into the earth. Or products can be 'technical nutrients' that continually circulate as pure and valuable materials within a closed-loop industrial cycle (rather than recycling). These products will also minimize indoor air pollution.

Bower (2000), author of *Healthy House Building for the New Millennium: A Design & Construction Guide,* is a house builder and the founder of the Healthy House Institute. He has created an in-depth guide to building a healthier house. Readers are taken step-by-step, through the construction of a safe, environmentally sound house. His methods can be used to increase the safety in new homes or in remodeling projects.

8.6.2 Trade magazines

Interiors and Sources, a trade magazine for the commercial interior design trade, is geared toward the designer who makes decision regarding products and services for commercial structures. *Interiors and Sources* is available as a traditional magazine and online. Online, the site includes a Green Design link and is the sponsor of EnvironDesign. More information may be obtained at http://interiorsandsources.com.

8.6.3 Research and interest groups

The Green Indoor Environment's website is a gateway to a variety of United States Environmental Protection Agency (EPA) programs that are designed to improve interior environments. Green Indoor Environment resources can be found at http://www.epa.gov/iaq/greenbuilding. Other countries have similar websites: United Kingdom, http://www.environment-agency.gov.uk; Australia, http://www.environment.gov.au/index.html; Denmark, http://glwww.mst.dk/homepage/.

The Environmental Design Research Association (EDRA) was created to advance and disseminate behavior and design research related to people and their environment. Researchers, educators, practitioners, and others are members of this organization who come together with a common interest in the environment. More information can be found at http://edra.org.

8.6.4 Websites

GREENGUARD Environmental Institute (GEI), an industry-independent non-profit organization, oversees the GREENGUARD Certification Program (Greenguard, n/d). Their mission is to improve public health and quality of life, and they set acceptable indoor air standards for various indoor products, environments, and buildings. Among the textile type products are bedding, textiles, furniture – seating, and wall finishes – wall fabrics. Under textiles, subcategories include draperies, panel fabrics, upholstery, and window shades. More information may be found at http://www.greenguard.org.

The United States Green Building Council (USGBC) promotes the construction and remodeling of 'buildings that are environmentally responsible, profitable and healthy places to live and work' (USGBC, 2007). Its *Leadership in Energy and Environmental Design* (LEED), a Green Building Rating System™, provides a benchmark for the design, construction, and operation of green buildings so that building performance can be assessed and sustainability goals met. LEED has devised a plan for measuring and documenting success for every building type and phase of a building lifecycle, which includes site development, energy efficiency, materials selection, water savings, and indoor environmental quality (USGBC, 2007). More information may be found at http://www.usgbc.org.

8.6.5 Social responsibility

It is important to select the appropriate textiles to minimize indoor pollution and, particularly, to promote good IAQ. Then, indoor environments will be safe for individuals with MCS, asthma, allergies as well as all occupants of the space (Bower, 2001; EPA, 1994). Particularly, an asthmatic attack can be caused by exposure to mould or mildew growth, tobacco smoke, and other indoor particles, and according to Bower (2001), 'there were 4580 asthma-related deaths in the United States in 1980'. Consumers and designers have a responsibility to demand, specify, and use materials that are environmentally safe and promote good IAQ. Clearly, there is no one solution for interior environments. Often the best solution is to determine what is appropriate for the individual or individuals who will occupy the space.

8.7 References

AIR QUALITY EDUCATION (2007a), 'Indoor air quality', *Kentucky Division for Air Quality*, Retrieved on June 25, 2007, from http://www.air.ky.gov/FAQ/Indoor+Air+Quality.htm

AIR QUALITY EDUCATION (2007b), Mould, *Kentucky Division for Air Quality*, Retrieved on June 25, 2007, from http://www.air.ky.gov/FAQ/Indoor+Air+Quality.htm

ALLEN P S, JONES L M and STIMPSON M F (2004), *Beginnings of interior environments*, (9th Ed), Upper Saddle River, New Jersey, Pearson Prentice Hall.

AMERICAN INSTITUTE OF ARCHITECTS – COLORADO (AIA) (2005), *Sustainable design resource guide,* Retrieved June 27, 2003, from http://www.aiacolorado.org/SDRG/div09/index.html

ANTRON CARPET FIBRE (2007), 'Reclamation and recycling – life after carpet', *Sustainability.* http://antron.net/content/sustainability/ant08_04_02.shtml

BINGGELI C (2007), *Interior design: A survey,* New York, John Wiley & Sons.

BIRNBAUM L S and STASKAL D F (2004), 'Brominated flame retardants: Cause for concern?' *Environmental Health Perspectives,* 112(1), 9–17.

BONDA P and SOSNOWCHIK K (2007), *Sustainable commercial interiors,* New York, John Wiley & Sons.

BOWER J (2000), *Healthy House Building for the New Millennium: A Design and Construction Guide*, Health House Institute, USA.

BOWER J (2001), *The healthy house: How to buy one, how to build, how to cure a sick one* (4th Ed), New York, Healthy House Institute.

BOWER L M (2007), 'Dry cleaning', *Healthy House Institute*, Retrieved June 30, 2007 from http://www.healthyhouseinstitute.com/a_733-Dry_Cleaning

CAPEK S and GROSE B (2005), *Environmentally responsible carpet choices*, National Parks Service – Pacific West Region, Retrieved on August 28, 2007 from, http://www.metrokc.gov/procure/green/carpet.htm#7

CARNEGIE (n/d). Climatex Lifecycle™. Retrieved June 27, 2007, from http://www.carnegiefabrics.com/

CARPET and RUG INSTITUTE (CRI) (n/d), 'Is there a fibre that will not mould?' *Frequently Asked Questions*, Retrieved on June 22, 2007, from http://www.carpet-rug.org

CONE J E and SHUSTERMAN D (1999), 'Health effects of indoor odourants', *Environmental Health Perspectives,* 95, 53–59.

COUNCIL FOR SCIENTIFIC and INDUSTRIAL RESEARCH (CSIR) (2007), 'Who are we?' *CSIR: Our Future through Science*. Retrieved on July 5, 2007, from http://www.csir.co.za/plsql/ptl0002/PTL0002_PGE100_LOOSE_CONTENT?LOOSE_PAGE_NO=7001678

CRYPTON® (2007), 'Crypton FAQs'. *Crypton Inc*, Retrieved on August 1, 2007, from http://www.cryptonfabric.com/crypton/about.do?atabId=6

CRYPTON® GREEN (2007), 'Crypton is committed to green and clean'. *Crypton Inc*, Retrieved on August 1, 2007, from http://www.cryptonfabric.com/crypton/about.do?wtabId=3

DESIGNTEX (2005). About the company. Retrieved June 27, 2007, from http://www.dtex.com/about/abt_cohist.htm

DUST FREE® (2006), '5 minute guide to indoor air quality: Why is indoor air quality such a hot topic?' *Indoor Air Quality Solutions Since 1982*. Retrieved on July 5, 2007, from http://www.dustfree.com/iaqgide.htm.

ENCYCLOPÆDIA BRITANNICA, 'FIBRE, MAN-MADE' (2007a), In *Encyclopedia Britannica*, Retrieved June 26, 2007, from Encyclopedia Britannica Online: http://www.britannica.com/eb/article-9108501

ENCYCLOPÆDIA BRITANNICA, 'SYNTHETIC FIBRE' (2007b), In *Encyclopedia Britannica*, Retrieved June 25, 2007, from Encyclopedia Britannica Online: http://www.britannica.com/eb/article-9070761

ENVIRONMENTAL BUILDING NEWS (EBN) (December 1994), 'Carpeting, Indoor Air Quality, and the Environment'. *Environmental Building News*, 3(6). Retrieved on July 1, 2007, from http://www.buildinggreen.com/auth/article.cfm?fileName=030601a.xml

ENVIRONMENTAL BUILDING NEWS (EBN) (June 2004), Flame Retardants Under Fire. *Environmental Building News*, 13(6). Retrieved on July 1, 2007, from http://www.buildinggreen.com/auth/article.cfm?fileName=130601a.xml

ENVIRONMENTAL PROTECTION AGENCY (EPA) (1994), *Indoor air pollution: An introduction for health professionals,* Retrieved March 3, 2003, from http://www.epa.gov/iaq/pubs/hpguide.html#faq1

ENVIRONMENTAL PROTECTION AGENCY (EPA) (1995), *The inside story: A guide to indoor air quality*. Retrieved March 3, 2003, from http://www.epa.gov/iaq/pubs/insidest.html.

ENVIRONMENTAL PROTECTION AGENCY (EPA) (2003), *Indoor air quality: Glossary of terms*. Retrieved March 3, 2003, from http://www.epa.gov/iaq/glossary.html

ENVIRONMENTAL PROTECTION AGENCY (EPA) (2007a), *An introduction to indoor air quality: Organic gases (volatile organic compounds – VOCs)*, Retrieved on June 22, 2007, from http://www.epa.gov/iaq/voc.html

ENVIRONMENTAL PROTECTION AGENCY (EPA) (2007b), *Appendix B: Introduction to Moulds,* Retrieved on June 22, 2007, from http://www.epa.gov/mould/append_b.html

FIBRES (2002), About our carpets: Fibres, *Environmental Home Center*, Retrieved on June 27, 2007, from http://www.environmentalhomecenter.com/learn.shtml?Directory_Code=topicsmain&Page_Code=carpets

FURNATURE (2003), 'Furnature: Healthier furniture for a healthier you', *Green Sleep® Furnature.* Retrieved on August 1, 2007, from http://furnature.com

GIST G (1999), 'Multiple chemical sensitivity: The role of environmental health professionals'. *Journal of Environmental Health*, 61(6) 4–6, Retrieved April 8, 2003, from InfoTRAC database.

GLASSWORKS (2003), 'Glassworks: Deepa Textiles', *The Journal of Design Architecture*, Retrieved on August 1, 2007 from http://www.designjournalmag.com/product/Glassworks.htm.

GODISH T (2001), *Indoor environmental quality*, New York: Lewis Publishers.

GREEN SEAL (2005), *About Green Seal*, Retrieved on May 20, 2007, from http://greenseal.org/about.htm

GREENGUARD (n/d), *About Greenguard*, Retrieved on May 20, 2007, from http://www.greenguard.org

HABERLE S E (n/d), *Go for green*, Retrieved June 26, 2003, from. http://www.asid.org/design_basics/design_specialities/sustaniable.asp

HEALTH & SAFETY (2006), 'FibreSource', *American Fibre Manufacturers Association*, Retrieved August 28, 2007, from http://www.fibresource.com/f-tutor/HEALTH.htm

IGIELSKA B, PECKA I, SITKO E, NIKEL G and WIGLUSZ R. (2002), Emission volatile organic compounds from new textile floor coverings, *Rocz Panstw Zakl Hig*, 53(3) 307–11.

IGIELSKA B, WIGLUSZ R, SITKO E and NIKEL G (2003), Release of volatile organic compounds from textile floor coverings in higher temperatures, *Rocz Panstw Zakl Hig*, 54(3) 329–35.

INDOOR AIR QUALITY (IAQ) (2006), 'Fibresource', *American Fibre Manufacturers Association*, Retrieved on June 30, 2007, from http://www.fibresource.com/f-tutor/HEALTH.htm

IRELAND J (2007), *Residential planning and design*, New York, Fairchild Books.

JACKA C (2001), 'Standards and responsibilities for green textile production in an emerging economy', *Pursuit*, Retrieved on June 27, 2007, from http://www.pursuit.co.za/archive/augsep_environment.htm

JACKMAN D, DIXON M and CONDRA J (2003), *The guide to textiles for interiors*, 3rd Ed, Winnipeg, Manitoba, Canada, Portage and Main Press.

KOSHIRO T (1974), A new flame retardant synthetic fibre made by emulsion spinning process, *Angewandte Makromolekulare Chemie*, 40(1) 277–290.

LANG S (2003), 'Waste fibre can be recycled into valuable products using new techniques of electrospinning', *Cornell News*, Retrieved on June 27, 2007 from http://www.news.cornell.edu/releases/Sept03/electrospinning.ACS.ssl.html

MATE K, ECO BALANCE SUSTAINABLE DESIGN CONSULTANCY, and MILNE G (2004), 'Material use: Indoor air quality', *Technical Manual for Life Style and the Future:*

Australia's Guild to Environmentally Sustainable Homes, Retrieved on June 22, 2007, from http://www.greenhouse.gov.au/yourhome/technical/fs33.htm

MCDONOUGH W and BRAUNGART M (2002a), *Cradle to cradle: Remaking the way we make things*, New York: Farrar, Straus and Giroux.

MCDONOUGH W and BRAUNGART M (2002b), *Transforming the textile industry*, Retrieved June 20, 2007, from http://www.mcdonough.com/writings/transforming_textile.htm

MILLER C S (1994), 'White paper: Chemical sensitivity: History and phenomenology', *Toxicology and Industrial Health*, 10(4/5) 253–276.

MINNESOTA DEPARTMENT OF HEALTH (MDH) (2007), 'Volatile organic compounds (VOCs) in your home', *Minnesota Department of Health*, Retrieved on June 22, 2007, from http://www.health.state.mn.us/divs/eh/indoorair/voc/

MONTANA STATE UNIVERSITY EXTENSION SERVICE (MSU-EXTENSION) (2007), 'Eliminate moulds, excessive moisture and other biological pollutants', *Healthy Indoor Air in American Homes*, Retrieved June 25, 2007, from http://www.montana.edu/wwwcxair/facts_mould.html

NIELSEN K (2007), *Interior textiles: Fabrics, application, and historical styles*, New York, Wiley.

NIVANA SAFE HAVEN (n/d), *Nivana safe haven – Creating healing, healthy environments*, Retrieved on August 1, 2007, from http://www.non-toxic.com.

NUSSBAUMER L (2004), Multiple chemical sensitivity: The controversy and relation to interior design, *Journal of Interior Design*, 30(2) 51–65.

OCCUPATIONAL SAFETY and HEALTH ADMINISTRATION (OSHA) (1994), 'Indoor air quality – 59:15968–16039', *U S Department of Labor*, Retrieved on July 5, 2007, from http://www.osha.gov/pls/oshaweb/owadisp.show_document?p_table=FEDERAL_REGISTER&p_id=13369

OCCUPATIONAL SAFETY and HEALTH ADMINISTRATION (OSHA) (2001), 'Indoor air quality – 66:64946', *U S Department of Labor*, Retrieved on July 5, 2007, from http://www.osha.gov/pls/oshaweb/owadisp.show_document?p_table=FEDERAL_REGISTER&p_id=16954

ORGANIC and HEALTHY (2007), 'Natural fibre non-toxic carpet & area rugs', *Organic and Healthy*, Retrieved on August 28, 2007, from http://www.organicandhealthy.com/carpet.html

PILATOWICZ G (1995), *Eco-interiors: A guide to environmentally conscious design,* New York, John Wiley & Sons.

REWI A J (n/d), 'Tactile Technology', *IIDA Knowledge Center*, Retrieved on June 27, 2007, from http://designmatters.net/features/0105tactile.htm

SATARA INC (n/d), 'Satara Inc: Home and baby store', *Satara Inc*, Retrieved on August 1, 2007, from http://www.satara-inc.com.

SCIENCE DAILY (2004), 'Household activities release a cloud of dust, increasing exposure to particulate pollution', *American Chemical Society*, Retrieved on June 26, 2007, from http://www.sciencedaily.com/releases/2004/03/040310080718.htm

SUITE SLEEP, (2007), 'Our products: Every ingredient is organic and green . . . naturally', *Suite Sleep*, Retrieved on August 1, 2007, from http://www.suitesleep.com/products.

THIVIERGE B (1999), 'Multiple chemical sensitivity', In *Gale Encyclopedia of Medicine,* (1st Ed) (pp. 1953), Toronto: Gale Research, Retrieved, April 20, 2003, from InfoTRAC database.

TOON J (2006), 'Assessing Indoor Air: New Test Facility Will Help Manufacturers Improve the Environment Inside Buildings', *Georgia Tech Research News,* Retrieved on July 3, 2007, from http://gtresearchnews.gatech.edu/newsrelease/envir-chamber.htm

U S GREEN BUILDING COUNCIL (2007), *Leadership in energy and environmental design (LEED)*, Retrieved on May 20, 2007, from http://www.usgbc.org

WINCHIP S M (2007), *Sustainable design for interior environments*, New York, Fairchild.

WITTENBERG J S (1996), *The rebellious body: Reclaim your life from environmental illness or chronic fatigue syndrome*, New York, Insight Books.

WOOL FIBRE (2007), *Australian Wool Innovation (AWI)*, Retrieved on August 28, 2007, from http://www.woolinnovation.com.au/Student_information/Wool_fibre/page__2157.aspx

WRIGHT S (2006), 'Healthy Indoor Air,' *The Old House Web*, Retrieved on June 25, 2007, from http://www.oldhouseweb.com/stories/Detailed/753.shtml

YEAGER J I and TETER-JUSTICE L K (2000), *Textiles for residential and commercial interiors* (2nd Ed), New York, Fairchild Publications.

ZHAO D, LITTLE J C and HODGSON A T (2002), 'Modeling the reversible, diffusive sink effect in response to transient contaminant sources', *Indoor Air*, 12 184–190.

9 Developments in flame retardants for interior materials and textiles

S. POSNER, Swerea IVF, Sweden

Abstract: This chapter discusses several types of flame retardant systems that are feasible for textiles. The main categories of chemical- and non-chemical flame retardant systems are discussed, so called barrier technologies, their basic mechanisms during fire, and compatibility with natural and polymeric textile materials. The chapter also includes a market review of and forecast information on the most common flame retardants on the market. Finally, developments of new flame retardant systems are described, especially designed for requirements of efficiency, health and environmental friendliness. One of the largest areas of research and development involves the use of nanotechnology to impart flame retardancy and increased functionality to textiles and other areas. However, technical and commercial viability is still limited and their future use in commercial settings remains questionable; only a few combinations of nano flame retardants with traditional flame retardants have met performance requirements.

Key words: flame retardants, additives, reactive, fillers, intumescent, smoke supressants, textiles, interior, fire mechanisms.

9.1 Introduction

Plastics are synthetic organic materials with a high content of carbon and hydrogen, which makes them combustible. For various applications in the building, electrical, transportation, mining and other industries, plastics have to fulfil flame retardancy regulatory requirements, mostly as mandatory specifications. Compliance with the flame retardancy requirements for plastics is controlled with numerous flammability tests. A large number of flame retardants have been evolved, due to these flammability requirements.

In contrast to most additives, flame retardants can appreciably impair the properties of plastics. The basic problem is the compromise between the decrease of performance caused by the flame retardant in the plastic and the flame retardant requirements.

An ideal flame retardant should:

- be easy to incorporate in the plastic involved
- be compatible with the plastic involved

- not alter the mechanical properties of the plastic
- be colourless
- have good light stability
- be resistant towards ageing and hydrolysis
- be extractable for recyclability of the polymer
- match and begin its thermal behaviour before the thermal decomposition of the plastic
- not cause corrosion
- not have harmful physiological effects
- not emit or, at least, emit low levels of toxic gases
- be as cheap as possible.

These qualities are, of course, impossible to reach for any single application. However, many formulations have been developed for each plastic in order to give as good properties as possible.

9.1.1 The global market for flame retardants

Table 9.1 gives the global consumption of flame retardants and their geographical distribution in 2005.

An approximate distribution of major categories of flame retardants produced world wide is described in Fig. 9.1.

Table 9.1 Consumption of flame retardants and their geographical distribution (2005)

Category	United States	Europe	Japan	Other Asian countries	Total volume (1000 metric tonnes)	Value (million US$)
Aluminium hydroxide	315	235	47	48	645	424
Organo phosphorus FRs	65	95	30	14	205	645
Brominated FRs	66	56	50	139	311	930
Antimony trioxide	33	22	17	44	115	523
Chlorinated FRs	33	35	5	10	82	146
Other FRs	51	47	11	14	123	197
TOTAL	**563**	**490**	**160**	**269**	**1481**	**2865**

Source: SRI Consulting.

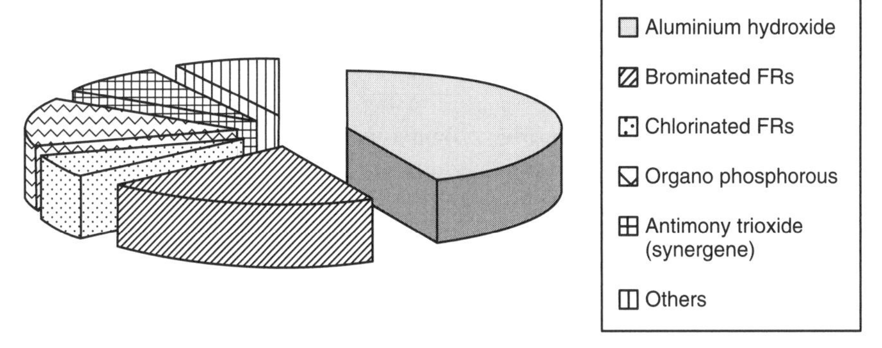

9.1 Global distribution (by weight) of major categories of flame retardants.
Source: SRI Consulting, 2005.

Most present-day flame retardants are used in the area of electronics in the manufacture of circuit boards and casings for home and office electronics, including mobile phone equipment. The plastics industry is by far the largest user of flame retardants, the largest quantities of which are supplied to raw-material manufacturers. A smaller proportion of world production of flame retardants goes to the textile and paper industries.

Brominated flame retardants have a significant share of the current global market. The production of halogenated flame retardants is equivalent to around 27% of total world production of flame retardants according to statistics from 2005. Brominated flame retardants account for about 21%. However, the predominant market share by volume is for the inorganic flame retardant aluminium hydroxide and to a lesser extent the organo phosphorous flame retardants.

The aromatic polybrominated flame retardants consist mainly of compounds with five, eight or ten bromine atoms in the structure. The most commonly used today is the compound containing ten bromine atoms, decabromodiphenyl ether (decaBDE), which belongs to the category of additive flame retardants. This means that the flame retardant is not chemically bound to the flame retardant material, unlike the reactive flame retardants. DecaBDE is the flame retardant produced in the greatest quantity around the world in terms of tonnes per year among the organic aromatic bromine compounds. Around 90% of world production ends up in electronics and plastics, while the other approximately 10% ends up in coated textiles and upholstered furniture and bedding products.

Tetrabrombisphenol A (TBBPA) and hexabromcyclododecane (HBCD) are two of the most common halogenated flame retardants used mainly in polymers for electronic devices and building insulation materials. There is a minor use for HBCD in coated textiles. In contrast to the additive flame retardant HBCD, TBBPA can be applied both as reactive and additive flame retardant depending of the kind of application to be manufactured. Around 70% of all TBBPA manufactured on the global market is applied as a reactive flame retardant. Since reactive flame-retardants are added during the polymerization process and become an integral part of the polymer, this results in a *modified* polymer with flame retardant properties, different in its molecule structure compared to the original polymer molecule. This prevents the reactive TBBPA from leaving the polymer and keeps the flame retardant properties intact over time with no emission to the environment.

Since additive flame-retardants are only *physically* bonded to the polymer as low molecular weight molecules, they may therefore be exposed and have a possible impact on the environment, depending on the original characteristics of the particular molecule. This is the case for HBCD and for the additive TBBPA.

TBBPA is used mainly in epoxy resins, both as a reactive and additive flame retardant. It also occurs at a lesser extent in high impact polystyrene (HIPS) and styrenics, which is a collective name for various styrene-based resins, such as acrylonitrile butadiene styrene (ABS), styrene–acrylonitrile (SAN), styrene butadiene rubber (SBR), styrene butadiene latexes (SBL) and unsaturated polyester resins (UPR).

There are a number of reactive nitrogen and phosphorous compounds, and inorganic alternatives, mainly alumina trihydrate, that can be used on a commercial basis as alternative non-halogen flame retardants in epoxy resins instead of TBBPA. For ABS, there is no obvious alternative to TBBPA. In the literature, boron salts and other inorganic flame retardants are described but there is also contradictory information as well. The market alternatives to ABS treated with TBBPA are combinations of polycarbonate (PC) and ABS, flame retarded with organic phosphorous compounds.

HBCD is applied mainly as an additive flame retardant in expanded polystyrene (EPS) used for insulation materials in the building and construction industry. This flame retardant is also used to a minor extent in HIPS for casings materials and in textile back coatings. Alternative non-halogen flame retardants exist on the market for EPS, mainly as organic and inorganic phosphorous compounds. There are also some examples described in the literature where boron and metal salts have been used. For textile coatings, barrier technologies such as intumescent systems are widely described as alternatives to HBCD.

9.2 General characteristics of flame retardant chemicals

Fire is a mechanism that occurs in three stages: (a) thermal decomposition whereby the solid (or condensed) phase breaks down to give gaseous decomposition products as a result of heat, (b) combustion chain reactions in the gas phase, where thermal decomposition products react with an oxidant (usually air) and generate more combustion products, which can then propagate the fire and release heat and (c) transfer of the heat generated from the combustion process back to the condensed phase to continue the thermal decomposition process (Hirschler, 1992; Beyler and Hirschler, 2002).

In general, flame retardants function by decreasing the likelihood of a fire occurring and/or decreasing the undesirable consequences of a fire (Lyons, 1970; Cullis and Hirschler, 1981). The action of flame retardants could result from raising the ignition temperature (although that is a rare occurrence, since flame retardants most often decrease the ignition temperature by acting in the gas phase (Hirschler, 1982)). Flame retardants usually decrease heat release rate (Hirschler, 1994), thus reducing the burning rate or flame spread of a fire, and/or by reducing smoke generation (Morose, 2005). Once thermal decomposition or ignition has occurred, flame retardants act in the gas phase by interfering with the chain reactions that cause the combustion process to propagate, and thereby reduce the tendency of the fire to spread. Alternatively, the flame retardants can act in the gas phase by cooling and thereby decreasing the rate of combustion. Other key mechanisms are condensed phase (or solid phase) activities whereby a solid charry (or glassy) layer is formed which interferes with the transfer of heat back from the gas phase to the condensed phase and inhibits (or prevents) further thermal decomposition.

In slightly more detail, the burning of a polymer is a very complex process involving a number of interrelated and interdependent stages. It is possible to decrease the overall rate of polymer combustion by interfering with one or more of these constituent stages. This is nearly always achieved by the use of additives, sometimes incorporated during the polymer synthesis, which may affect the various stages of the burning process in a number of different ways. The simplest way, in theory, of preventing polymer combustion is to design the polymer so that it is thermally very stable. Its initial breakdown will be effectively prevented and the whole combustion process will never be able to be initiated with that polymer. Thermally stable polymers are usually more difficult to process and do not possess many of the other performance characteristics needed for wide acceptance. Furthermore, they are often very expensive to produce. This approach provides no help in imparting flame retardance to the organic polymers in widespread use.

The additives which are widely used in practice to inhibit the combustion of the commoner synthetic organic polymers are generally incorporated throughout the polymeric material, although they can also be coated on the external surface of the polymer to form a suitable protective barrier. Thus, flame retardants can be classified, broadly speaking, into two types:

(i) Reactive flame retardants are used by modifying the production of existing polymers by replacement of one or more of the unsubstituted reactant monomers normally used, by a substituted monomer containing heteroelements known to confer some degree of flame retardance. The substituted monomers are then known as *reactive* flame retardants and the heteroelements become an integral part of the resulting polymer structure. Compounds containing such elements can also be introduced into the monomers during the polymerization process in such a way that they chemically alter occasional structural units in the polymer in the same way as a comonomer. Such additives also constitute reactive flame retardants. Reactive flame retardants need to be incorporated at an early stage of manufacture but, once introduced, become a permanent part of the polymer structure. Once chemically bound, the reactive flame retardants chemicals cease to exist as separate chemical entities. It has been stated that reactive flame retardants can have a greater effect than additive flame retardants on the chemical and physical properties of the polymer into which they are incorporated (U.S. EPA, 2005).

(ii) Alternatively, compounds containing elements with known flame-retardant properties, can simply be intimately mixed with existing polymers in order to make them less susceptible to combustion. In this case, the compounds do not react chemically with the polymers and are known as *additive* flame retardants. From the manufacturing point of view, the introduction of additive flame retardants is usually the simplest way of improving the fire performance of an organic polymer because the additives can be incorporated during the final stages of production of the finished material.

Combinations of flame retardants are often used which are said to have synergistic or antagonistic effects. Synergism is a larger-than-additive effect, whereas antagonism is a smaller-than-additive effect. The use of the term additive here is generally based on the assumption that the effects of flame retardants vary linearly with their concentrations. This is not, however, usually the case (Weil, 1975, 2000). Several combinations of flame retardants have therefore been misleadingly termed synergistic when the 'other' flame retardant simply has a beneficial effect. Sometimes a heteroatom already present in the polymer backbone may interact with a flame retardant and thus exhibit synergism or antagonism.

Both reactive and additive flame retardants often have deleterious effects on the properties of the polymers into which they are incorporated. Thus, they may change the viscosity, flexibility, and density, and may also increase the susceptibility of the polymers to photochemical and thermal degradation. Sometimes they have disastrous effects on electrical properties, increasing the electrical conductivity to unacceptable levels.

Typically, flame retardants contain one of the following seven elements: chlorine, bromine (both halogens), aluminum, boron, nitrogen, phosphorus or antimony (Lyons, 1970, Cullis and Hirschler, 1981; Hirschler, 1982). There are, however, a number of replacements and synergists that have also been shown to be effective. For example, aluminum, which is most often used as an oxide or hydroxide, can be replaced by magnesium hydroxide or by a magnesium salt. Also, some elements, such as zinc, often used as zinc borate or zinc stannate, and molybdenum, often used as ammonium molybdates, have also been found to be effective, primarily as smoke suppressants in mixtures of flame retardants.

The most obvious stage that can be affected by an additive is the initial thermal decomposition of the polymer to give combustible gaseous products. In other words, a flame retardant may alter the breakdown of the polymer in such a way that either the nature or the rate of evolution of the gaseous decomposition products is changed. In consequence, when these products mix with the oxidant, the composition of the resulting mixture may no longer lie within the flammable range. There are, however, several different mechanisms by which the breakdown of the solid polymer can be affected. Clearly, in some cases, a flame retardant acts simply by chemical termination in the condensed phase of the free-radical chains by which thermal decomposition of the polymer takes place. Many flame retardants affect this initial stage of the process, more often than not by accelerating thermal decomposition.

A completely different mode of action is that exerted by apparently inert solids incorporated into polymers. Such materials, which are generally known as fillers, tend, when present in considerable quantities, both to act as heat sinks as a result of their heat capacity and to conduct heat away by virtue of their thermal conductivity; they may thus prevent polymers from reaching temperatures at which they undergo significant breakdown to yield combustible gaseous products. The temperature is, of course, kept down even more effectively if compounds are used which decompose endothermically. This is a mechanism of action of some fillers, including metal oxides and hydroxides. As the mechanism of action of these flame retardants is, to a large extent, a physical rather than a chemical effect, large levels of the additives are needed.

Another, quite distinct, method of imparting flame retardance is to arrange that the surface of the polymer is or becomes covered with a non-flammable protective coating. This will help to insulate the flammable polymer from the source of heat and thus prevent the formation, or at any rate the escape into the gas phase, of the combustible breakdown products; it may also exclude the gaseous oxidant, normally air or oxygen, from the surface of the polymer. Certain compounds which may be initially incorporated into or simply coated on an organic polymer either decompose or react with other materials in the condensed phase at high temperatures to give a protective barrier in which the gaseous products of polymer decomposition are trapped as they are formed. An *intumescent* coating is then said to have been formed on the polymer surface. Alternatively, direct application of a non-flammable layer on the surface of the polymer yields a non-intumescent coating. This is a typical mechanism of action of many phosphorus-containing species.

The other stage in the burning of an organic polymer which can clearly be inhibited by additives is the combustion of the gaseous products of polymer decomposition or, in other words, the gaseous flame reactions. Here again there are several rather distinct possible modes of action. For example, a flame retardant may cause a reactive gaseous compound to be released into the combustion zone where it enters chemically into the flame reactions and reduces the flame speed, generally as a result of the replacement of highly active free radicals by less reactive ones. In other cases, the additive may cause the evolution during combustion of the polymer of a 'mist' of small particles which interferes with flame propagation by acting as 'third bodies' to catalyze free-radical recombination and hence chain termination. This is a typical mechanism of action of halogenated flame retardants. They usually act by decomposing and generating hydrogen chloride (or hydrogen bromide), which then reacts with oxygenated radicals and inhibits the combustion reactions in the gas phase (Cullis and Hirschler 1981, Hirschler 1982, Georlette *et al.*, 2000). Neither chlorine nor bromine is produced in this way as the halogen atoms will react much more quickly with hydrogen atoms or oxygenated species than with another halogen atom. Antimony oxide aids in the thermal decomposition and volatilization of halogenated additives (Hastie and McBee 1975), often by generating intermediate volatile species such as antimony oxyhalides.

Another possible mechanism of flame inhibition is for the additive to release, on decomposition, relatively large quantities of an inert gas, which changes the composition and perhaps also the temperature of the gaseous products of polymer decomposition so that the resulting mixture with the surrounding gaseous oxidant is no longer capable of flame propagation. In some systems, when the polymer burns, the flame-retardant additive appears to be released chemically unchanged and in the form of a heavy vapour which effectively 'smothers' the flame by interfering with the normal interchange of combustible gaseous polymer decomposition products and combustion air or oxygen. This is the typical mechanism of action of metal hydroxides, such as aluminum or magnesium hydroxide (Horn, 2000).

However, it is sometimes possible to inhibit combustion by interfering with the normal transfer of the heat of combustion back to the polymer. Thus, for example, certain additives may promote depolymerization and hence, as a result of the lowering of the molecular weight of the polymer, facilitate melting. As the burning melt drips away from the bulk of the polymer, it carries with it a substantial proportion of the heat which would otherwise be used to bring about decomposition of the polymer to volatile products and hence feed the flame. Thus, in principle, ready melting decreases flammability, although in practice droplets of burning molten polymer may help a fire to spread to other combustible materials.

Another method of delaying, if not preventing, combustion which is associated with heat transfer effects, is to coat or construct the polymer in such a way that, when it burns, incandescent sections disintegrate from the original polymer and thus remove much of the associated heat with them from the combustion zone. This mechanism of action, known as ablation, is in a sense the parallel, in the solid phase, of melting and dripping in the liquid phase. A surface char layer is frequently formed which isolates the bulk of the polymer material from the high temperature environment. This charry layer remains attached to the substrate for at least a short period while a degradation zone is formed underneath it. It is in this zone that the organic polymer undergoes melting, vaporization, oxidation or pyrolysis. The ablative performance of any polymeric material is influenced not only by its composition and structure (e.g. higher hydrogen, nitrogen, and oxygen contents of the polymer increase the char oxidation rate, while this rate decreases with increasing carbon content) but also by its environment (e.g. the oxygen content of the atmosphere) (Levchik and Wilkie, 2000).

Smoldering combustion, and the closely related phenomenon of glowing combustion, occurs principally with polymeric materials of high surface area which break down during combustion to form a residual carbonaceous char (typically cellulosic materials). Although it is not possible, in practice, to identify separate well-defined stages in the non-flaming combustion of organic polymers, the first stage conceptually is thermal or oxidative breakdown to form only relatively small proportions of volatile products and a solid char. The second stage consists of the ignition and burning of this char. Under conditions of smoldering, ignition of the carbonaceous residue may occur at temperatures well below the ignition temperature of the volatile products (Browne, 1958). Thus the first material to ignite is the char and the volatile products often do not burn, either because there is no local ignition source of sufficient intensity or because their rate of production is so low that the gaseous mixture formed is too lean. In principle, then, it is possible to inhibit non-flaming combustion either by retarding or preventing the initial breakdown of the polymer to form a char or by interfering with the further combustion of this char. Flame retardants which affect polymer decomposition must act in the condensed phase. The possible mechanisms are clearly the same as those that apply in the case of flaming combustion; chemical inhibitors, inert fillers, and protective coatings are all able to slow down or even prevent completely such decomposition. On the other hand, additives which interfere with the combustion of the carbonaceous char can do so by virtue of their effects on reactions taking place either in the solid phase or in the surrounding gas phase. Studies of the combustion of substantially pure carbons have shown that almost every impurity exhibits either a promoting or an inhibiting effect under appropriate experimental conditions. As far as organic polymers are concerned, however, the normal flame retardants used to prevent flaming combustion appear to have no effect on smouldering combustion. In general, only boric acid and phosphates are useful for preventing after-glow. There is little evidence that these compounds act by producing a glass-like surface coating which prevents access of the gaseous oxidant to the polymer. Instead, they probably interfere with the solid-phase reactions which lead to the formation of carbon oxides, inhibiting either the initial production of surface oxides or the highly exothermic conversion of carbon monoxide to carbon dioxide.

Another way in which flame retardants have been classified is into five main categories (Posner, 2008 and Morose, 2005):

(i) *Inorganic.* This category includes metal hydroxides (e.g. aluminum hydroxide and magnesium hydroxide), antimony compounds (e.g. antimony trioxide), boron compounds (e.g., zinc borate), and other metal compounds (molybdenum trioxide). These act by various different mechanisms and are probably best addressed individually. As a group, these flame retardants represent the largest fraction of total flame retardants in use.

(ii) *Halogenated.* These flame retardants are based primarily on chlorine and bromine. Typical halogenated flame retardants are halogenated paraffins, halogenated alicyclic and aromatic compounds and halogenated polymeric materials. Halogenated flame retardants also often contain other heteroelements, such as phosphorus or nitrogen. When antimony oxide is used, it is almost invariably as a synergist for halogenated flame retardants, ever since it was first used by the US army in the 1940s (Little, 1947). The effectiveness of halogenated additives, as stated previously, is due to their interference with the radical chain mechanism in the combustion process of the gas phase.

Brominated compounds represent approximately 25% by volume of the global flame retardant production (Morose, 2005). Chemically, they can be further divided into three classes:

- Aromatic, including tetrabromobisphenol A (TBBPA), polybrominated diphenyl ethers (PBDEs), and polybrominated biphenyls (PBBs).
- Aliphatic
- Cycloaliphatic, including hexabromocyclododecane (HBCD).

(iii) *Phosphorus-based.* This category represents about 20% by volume of the global production of flame retardants and includes organic and inorganic phosphates, phosphonates and phosphinates, as well as red phosphorus, thus covering a wide range of phosphorus compounds with different oxidation states. There are also halogenated phosphate esters, often used as flame retardant plasticizers but not commonly used in electronics applications (Hirschler 1998, Green, 2000, Weil 2004).

(iv) *Nitrogen-based.* These flame retardants include melamine and melamine derivatives (e.g. melamine cyanurate, melamine polyphosphate). It is rare for flame retardants to contain no heteroatom other than nitrogen and to be used on their own. Nitrogen-containing flame retardants are often used in combination with phosphorus-based flame retardants, usually with both elements in the same molecule.

(v) *Barrier technologies* (i.e intumescent systems). Intumescent (or swelling) systems have existed since the 1940s, principally in paints. Several intumescent systems linked to textile applications have been on the market for about 20 years, and have successfully shown their great potential. Intumescent systems include use of expandable graphite impregnated foams, surface treatments and barrier technologies of polymer materials.

The basic mechanisms of flame retardancy will vary, depending on the flame retardant and polymer system. Due to the differing physical and chemical properties of flame retardant chemicals, most are used exclusively as either reactive or additive flame retardants.

9.3 Flame retardants used in textiles

Like any other additives, a flame retardant will be selected for the particular properties it imparts to make it satisfy the specifications for the final compound established by the customer. As mentioned earlier, different flame retardants may be chosen to give different levels of fire protection, depending on the specific levels defined by the customer and that particular

Table 9.2 Overview of flame retardant systems applied in common commercial textiles applications (Posner, 2008)

Common commercial textile applications	Inorganic flame retardants	Phosphorous/nitrogen organic	Halogen organic flame retardants	Non chemical flame retardant systems and materials
Back coatings and impregnation for carpets, automotive seating, furniture in homes and public buildings, aircraft, underground.	Aluminium hydroxide Magnesium hydroxide Ammonium compounds (unspecified) Borax	Tetrakis hydroxiymethyl phosphonium salts such as chloride (THCP) or ammonium (THPX) Dimethyl phosphono (N-methylol) propionamide Diguanidine hydrogen phosphate Aromatic phosphates (unspecified) Dimethyl hydrogen phosphite (DMHP) Melamine (nitrogen based) Phospho nitrilic chloride (PNC) DOPO=Dihydrooxaphosphaphenanthrene oxide	Trichloropropyl phosphate HBCD PBDEs TCPP	Intumescent systems Aramide fibres (certain protective applications) Wool Modacrylic

market. New flame retardant solutions are constantly introduced and some disappear from the market for a number of reasons. Therefore Table 9.2 is an on-the-spot account and cannot be complete, but only act as a guide that illustrates the variety and optional chemical systems that are available and actually work as viable flame retardant systems for textiles. However, it needs to be clearly understood that each flame retardant application is specific and unique, and there is no single universal solution for fire protection of materials and applications.

9.3.1 Aryl phosphates

This large group of organophosphorus flame retardants include triphenyl, isopropyl – and t-butylsubstituted triaryl and cresyl phosphates. Phosphates with larger substitution carbon chains (therefore less volatile) are commercially available beside those mentioned above.

Aryl phosphates are used as flame retardant for plasticizers PVC. It has been shown that although PVC does not require any flame retardancy as a polymer, the addition of phthalate plasticizers makes PVC flammable. Triaryl phosphates are more efficient flame retardants than the alkylated triaryl phosphates. However, the alkylated triaryl phosphates are shown to be more efficient plasticizers than triaryl phosphates.

9.3.2 Halogen containing phosphorus flame retardants

Chloro alkyl phosphates have been found effective in flexible polyurethane (PUR) foams, but since they are not stable during curing reactions of PUR, which is a strong highly exothermic reaction (with heat generated), they cause discolouring problems.

9.3.3 DOPO

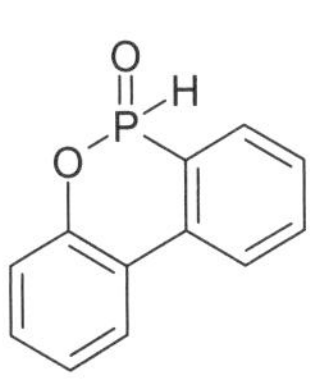

O
P
H
O

DOPO is a hydrogenphosphinate made from *o*-phenyphenol and phosphorus trichloride and was originally developed as a flame retardant for polyester textile fibers. It also has an application as an antioxidant-type stabilizer

(Weil and Levchik, 2004). To decrease the cost of their formulations, some laminate manufacturers are using DOPO in combination with less expensive materials such as ATH and/or silica (Thomas *et al.*, 2005) or along with more cost-effective compounds such as metal phosphinates (Clariant, 2005).

9.4 Flame retardant fillers

9.4.1 Aluminum hydroxide

Aluminum hydroxide ($Al(OH)_3$) is the largest volume flame retardant used worldwide, with an estimated 42% volume market share in 2006 (BCC, 2006). Aluminum hydroxide is commonly referred to as alumina trihydrate (ATH) and is currently used to impart flame retardancy and smoke suppression in carpet backing, rubber products, fiberglass-reinforced polyesters, cables, and other products. It is also used in the manufacture of a variety of items – antiperspirants, toothpaste, detergents, paper, and prinking inks – and is used as an antacid.

When heated to 200–220 °C, ATH begins to undergo an endothermic decomposition to 66% alumina and 34% water (Morose, 2005). It retards the combustion of polymers by acting as a 'heat sink' – i.e. by absorbing a large portion of the heat of combustion (HDPUG, 2004).

The addition of large loadings of ATH (or of magnesium hydroxide) is often made easier by using organosilicon-compounds as processing aids, which provide the necessary dispersibility and processability of the fillers in polymeric compositions.

9.4.2 Magnesium hydroxide

Magnesium hydroxide acts, in general, the same way as ATH, but it thermally decomposes at slightly higher temperatures (around 325 °C). Combinations of ATH and magnesium hydroxide function as very efficient smoke suppressants in PVC.

9.4.3 Melamine polyphosphate

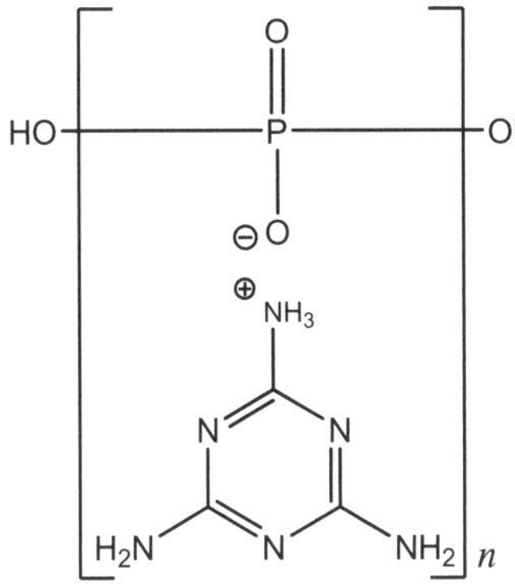

Melamine polyphosphate, a nitrogen-based flame retardant, is typically used as crystalline powder and in combination with phosphorus-based compounds. Its volume market share in 2006 was slightly more than 1% (BCC, 2006) but is expected to increase as the demand for halogen-free alternatives increases. Similar to ATH, melamine polyphosphate undergoes endothermic decomposition but at a higher temperature (350°C). It retards combustion when the released phosphoric acid coats, and therefore forms a char around the polymer, thus reducing the amount of oxygen present at the combustion source (Special Chem, 2007).

9.4.4 Silica

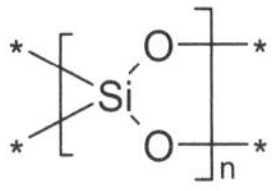

Also known as silicon dioxide (SiO_2), silica is characterized by its abrasion resistance, electrical insulation, and high thermal stability. Silica is not a flame retardant in the traditional sense. It dilutes the mass of combustible components, thus reducing the amount of FR necessary to pass the flammability test. Silica is most commonly used in combination with novolac-type epoxy resins. For example, silica clusters can be reacted with phenolic novolac resins (the resin bonds to hydroxyl groups on the silica cluster) to form a silica–novolac hybrid resin (US Patent Storm, 2002). During combustion, silica accumulates near the polymer surface to insulate the rest of the material, decreasing the heat release and mass loss rate. Unlike most other fillers, silica migrates to the polymer surface via physical processes instead of chemical reactions.

9.4.5 Ammonium polyphosphate

Ammonium polyphosphate (APP) is used mainly as an acid source in intumescent systems. However, APP alone is an effective flame retardant for polyamides and similar polymers (Posner, 2008).

9.4.6 Antimony oxide

Antimony oxide, typically antimony trioxide (Sb_2O_3), can be used as a flame retardant in a wide range of plastics, rubbers, paper and textiles. Antimony trioxide does not usually act directly as a flame retardant, but as a synergist for halogenated flame retardants. Antimony trioxide enhances the activity of halogenated flame retardants by releasing the halogenated radicals in a stepwise manner. This retards gas-phase chain reactions associated with combustion, which slows fire spread (Hastie and McBee 1975, Hirschler, 1982, Chemical land 21, 2007).

9.4.7 Melamine cyanurate

Melamine cyanurate is relatively cheap and highly available. However, it is a poor flame retardant and requires high dosage (>40% weight) (Albemarle, 2007).

9.5 Future developments in flame retardant chemicals

One of the largest areas of research and development involves the use of nanotechnology to impart flame retardancy and increased functionality to textiles and other products. However, their technical and commercial viability is still limited and their future use in commercial settings remains questionable. So far, only combinations of nano flame retardants with traditional flame retardants have met performance requirements. In addition, these new nano–traditional flame retardant combinations are usable only in certain polymer systems.

One type of halogen-free nano flame retardant is being developed through the synthesis of ethylene–vinyl acetate (EVA) copolymers with nanofillers (or nanocomposites) made of modified layered silicates (Beyer, 2005). Nanofillers are incorporated into the polymer during the polymerization process by treating the surface of the nanofiller to induce hydrophobic tendencies. The hydrophobic nanofiller disperses in the monomers, which then undergo polymerization and trap the nanofillers. (Nanocor, 2007). Nanocomposites can also incorporate aluminum into their structures, and can be combined with additive flame retardants, such as aluminum hydroxide (ATH), leading to a reduction of the total ATH content and a corresponding improvement in mechanical properties (Beyer, 2005). Unfortunately, the use of nanofillers has been attempted on a number of different polymers but has only been able to be commercialized to date on the EVA cable insulating materials developed by Beyer.

9.6 References

ALBERMARLE. The Future Regulatory Landscape of Flame Retardants from an Industry Perspective. In *Environmentally Friendly Flame Retardants*, Proceedings of the Intertech, Pira Conference, Baltimore, MD, July 19, 2007.

BCC RESEARCH. *Flame Retardancy News* **2006,** *16* (3).

BEYER, GUNTER. Flame Retardancy of Nanocomposites – from Research to Technical Products. *J. Fire Sci.* **2005,** *23* (Jan).

BEYLER, C.L. and M. M. HIRSCHLER, 'Thermal Decomposition of Polymers', Chapters 1–7 in *SFPE Handbook of Fire Protection Engineering* (3rd Edition), Editor-in-chief: P.J. DiNenno, pp. 1/110–1/131, NFPA, Quincy, MA, 2002.

CHEMICAL LAND 21. Antimony Oxide. http://www.chemicalland21.com/arokorhi/industrialchem/inorganic/ANTIMONYTRIOXIDE.htm (accessed 2007).

CULLIS, C.F. and M.M. HIRSCHLER, *The Combustion of Organic Polymers*, Oxford University Press, Oxford, 1981.

GEORLETTE, P., J. SIMONS and L. COSTA, 'Halogen-containing fire retardant compounds' Chapter 8 in *Fire Retardancy of Polymeric Materials*, Eds. A.F. Grand and C.A. Wilkie, p. 245, Marcel Dekker, New York, 2000.

GREEN, J., 'Phosphorus-containing flame retardants' Chapter 5 in *Fire Retardancy of Polymeric Materials*, Eds. A.F. Grand and C.A. Wilkie, p. 147, Marcel Dekker, New York, 2000.

HASTIE, J.W. and C.L. MCBEE, in *Halogenated fire suppressants*, (ed. R.G. Gann), American Chemical Society, Symposium Series # 16, American Chemical Society, Washington, DC., 1975.

HIRSCHLER, M.M., 'Recent developments in flame-retardant mechanisms', in *Developments in Polymer Stabilisation*, Vol. 5, Ed. G. Scott, pp. 107–52, Applied Science Publ., London, 1982.

HIRSCHLER, M.M. editor, *Fire hazard and fire risk assessment*, ASTM STP 1150, Amer. Soc. Testing and Materials, Philadelphia, PA, (1992).

HIRSCHLER, M.M., 'Fire Retardance, Smoke Toxicity and Fire Hazard', in *Proc. Flame Retardants '94*, British Plastics Federation Editor, Interscience Communications, London, UK, Jan. 26–27, 1994, pp. 225–37 (1994).

HIRSCHLER, M.M. 'Fire Performance of Poly(Vinyl Chloride) – Update and Recent Developments', *Flame Retardants '98*, February 3–4, 1998, London, pp. 103–23, Interscience Communications, London, UK, 1998.

HORN, W.E. JR, 'Inorganic hydroxides and hydroxycarbonates: Their function and use as flame-retardant additives', Chapter 9 in *Fire Retardancy of Polymeric Materials*, Eds. A.F. Grand and C.A. Wilkie, p. 285, Marcel Dekker, New York, 2000.

LEVCHIK, S. and C.A, WILKIE, 'Char formation', Chapter 6 in *Fire Retardancy of Polymeric Materials*, Eds. A.F. Grand and C.A. Wilkie, p. 171, Marcel Dekker, New York, 2000.

LITTLE, R.W. in *Flameproofing Textile Fabrics*, American Chemical Society, Monograph Series # 104, p. 248, Ed. R.W. Little, American Chemical Society, Washington, DC, (1947).

LYONS, J.W., *The Chemistry and Use of Fire Retardants*, Wiley, New York, 1970.

MOROSE, G. *An Investigation of Alternatives to Tetrabromobisphenol A (TBBPA) and Hexabromocyclododecane (HBCD).* Lowell Center for Sustainable Production: University of Massachusetts Lowell, 2006. Prepared for: The Jennifer Altman Foundation.

'NANOCOR. Nanomer nanoclay as flame retardation additives.' In *Environmentally Friendly Flame Retardants, Proceedings of the Intertech Pira Conference*, Baltimore, MD July 20, 2007.

POSNER, S., "Guidance *on flame-retardant alternatives to pentabromodiphenylether (PentaBDE)*" , UNEP , Persistant Organic Pollutants Review Committee, fourth meeting. on POPs for the Stockholm Convention, on behalf of the Norwegian pollution control authority (SFT), (2008).

SPECIAL CHEM. FLAME RETARDANTS CENTER: Melamine Compounds http://www.specialchem4polymers.com/tc/Melamine-Flame-Retardants/index.aspx?id=4004 (accessed 2007).

THOMAS, S. G., JR., M.L., HARDY, K.A., MAXWELL, P.F. RANKEN. 'Tetrabromobisphenol-A Versus, Alternatives in PWBs'. *OnBoard Technology* **2005,** (June).

US PATENT STORM, 2002. Polyphosphate salt of a 1, 3, 5-triazine compound with a high degree of condensation, a process for its preparation and use as flame retardant in polymer compositions. [Online], U.S. Patent Number 6369137, 2002. http://www.patentstorm.us/patents/6369137-description.html (accessed 2007).

WEIL, E.D. in *Flame Retardancy of Polymeric Materials*, Vol. 3, Eds. W.C. Kuryla and A.J. Papa, p. 185, Marcel Dekker, New York, 1975.

WEIL, E.D. 'Synergists, adjuvants and antagonists in flame-retardant systems', Chapter 4 in *Fire Retardancy of Polymeric Materials*, Eds. A.F. Grand and C.A. Wilkie, p. 115, Marcel Dekker, New York, 2000.

WEIL, E.D. and S. LEVCHIK, 'A Review of Current Flame Retardant Systems for Epoxy Resins'. *J. Fire Sci.* **2004,** 22 (Jan).

WEIL, E.D. 'Flame Retardants – Phosphorus Compounds,' in *Kirk-Othmer Encyclopedia of Chemical Technology*, John Wiley & Sons, Inc., NY, 1994, revised 2004.

10
Fire testing of upholstered furniture – current and future methods

C. M. FLEISCHMANN,
University of Canterbury, New Zealand

Abstract: This chapter discusses the fire behaviour of upholstered furniture and mattresses including how the fire evolves and how a change in materials used can affect combustion behaviour. Various mandatory and voluntary standard test methods for furniture and mattresses are described and the results of the latest research and enhanced fire modelling techniques are discussed.

Key words: soft combustibles, combustion behaviour, upholstered furniture, mattresses, fire modelling, fire testing.

10.1 Introduction

When an unwanted fire occurs within a residential dwelling, there are few objects that have the potential to bring about untenable conditions as swiftly as upholstered furniture. In the worst case scenario, the Heat Release Rate (HRR) of an upholstered furniture item can reach values of 3 MW in a very short period of time (3–5 minutes) following ignition.[1] Furthermore, it is not only the heat given off during the growth period, but also the toxic combustion products (primarily CO) produced that overwhelm the occupants. These coupled effects are directly linked to the materials used in the construction of the furniture, and in particular the soft combustibles (covering and padding materials).

This chapter describes the current level of understanding of the fire behaviour of upholstered furniture and mattresses. The first section will briefly describe the US fire statistics, which clearly show why we need research on the combustion behaviour of upholstered furniture and regulation to reduce the hazards to occupants. We will proceed with a description of how upholstered chairs burn once the fire is greater than about 20 kW. This description is intended to give the reader a better understanding of how a burning chair evolves and how changing the foam and fabric can make a significant difference in the combustion behaviour. Currently, there are several mandatory and voluntary standard test methods available for furniture and mattresses. The most common methods will be briefly described, with the appropriate references provided for further investigation by the reader. The ongoing work by the Consumer Product Safety Commission (CPSC) on regulating upholstered furniture and mattresses will also be summarized in this section.

Over the last decade there has been a great deal of research on upholstered furniture and mattresses, with support from the European Commission. The study Combustion Behavior of Upholstered Furniture (CBUF) incorporated both extensive experimental research as well as enhanced fire modelling. The results of this research are the basis for the discussion on modelling upholstered furniture fires in this chapter.

Finally the chapter will conclude with a brief section highlighting where additional information can be found.

10.2 The role of upholstered furniture and mattresses in fires

In the United States, between 2003 and 2005 inclusive, there were 8220 fire deaths in 1 135 900 residential structure fires from non-incendiary and non-suspicious fires.[2] This is an average of 2970 fire deaths/year. In approximately 2/3 of these fatal fires, the first material ignited was reported; in the remaining 1/3, the first item ignited went unreported. Figure 10.1 shows a breakdown of the average number of fire deaths per year versus the first material ignited based on the CPSC estimates. In 32% of these fatal fires, the first material ignited was upholstered furniture, and in

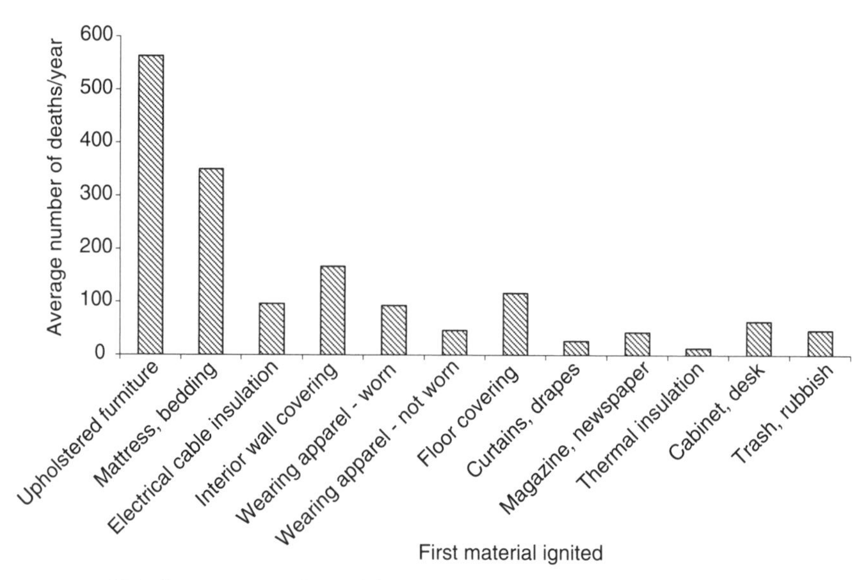

10.1 Average number of fatalities in residential fires (2003–05) broken down by first item ignited.

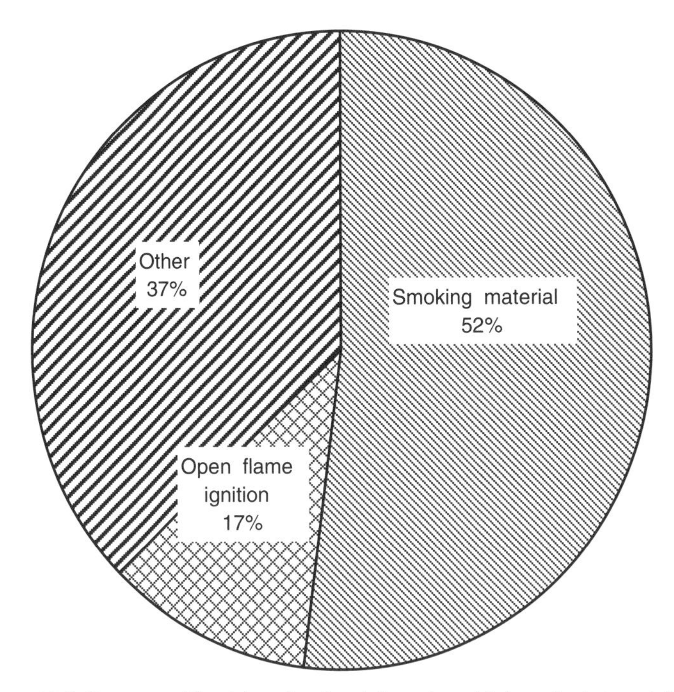

10.2 Source of ignition for fatal fires in which upholstered furniture or bedding/mattress was the first item ignited.

25% of the fires, the first material ignited was mattress/bedding. This clearly shows the important role upholstered furniture and mattress/bedding play in fatal fires.

In addition to the total number of fires, the individual ignition scenarios are also of interest when assessing the hazards associated with upholstered furniture and mattress/bedding fires. Research has shown that the fire hazard of upholstered furniture and mattress/bedding fires is a function of the ignition source. Figure 10.2 shows that smoking material is clearly the most common ignition source when upholstered furniture or mattress/bedding are the first items ignited, although open flames and other sources cannot be ignored. Clearly, upholstered furniture and mattress/bedding fires play a significant role in the fatal fire statistics, and further research and potentially more regulation is warranted to reduce the hazard posed by upholstered furniture and mattress/bedding fires.

10.3 Description of the burning behaviour of upholstered furniture[3]

This section discusses the results of a series of experiments conducted on domestic upholstered furniture. In this study, fifty-five 1- and 2-seater chairs, modelled after the CBUF standard chair, were burned. The following description is based on the ignition of an upholstered chair from a 30 kW ignition burner centred between the armrests, 50 mm from the back cushion, and placed 25 mm above the seat cushion, in accordance with the standard CAL TB133.[4] The burner is left on for 120 s and then removed. Previous research has shown that the location and intensity of the ignition source has little impact on the peak heat release rate; however, the ignitability and time to peak are greatly affected.[5] The application of the description given here is appropriate once radiation becomes the dominant form of heat transfer governing the burning rate. Thus, the ignition scenario is not included in this description.

A sophisticated load cell system was used to isolate the mass of the burning chair from the mass of the burning foam/fabric pool that collects on the platform below the chair. The burning chair was supported on an insulated steel frame, suspended 25 mm above the platform. The frame was physically isolated from the platform and supported on two load cells to continuously record the mass of the chair. The platform below the chair was supported on four load cells, one on each corner, to record the weight of the foam and fabric that fell from the chair. The heat release rate from the burning chair was measured using oxygen depletion calorimetery.

After flaming ignition has occurred, the general burning behaviour of an upholstered furniture item, ignited on the seat, can be broken down into four main phases: Spread; Burn Through; Pool Fire; and Burnout. Figure 10.3 shows the mass and Heat Release Rate (HRR) histories for a chair with domestic polyurethane foam covered by wool fabric. The four phases of burning have been highlighted on the plot.

Spread phase. Named for the flame spreading over the surfaces of the seat cushion, up the back cushion and onto the armrest assemblies. Figure 10.4 is a series of photographs taken during the same experiment shown in Fig. 10.3, i.e. domestic foam covered with wool fabric. The clock seen in the images is reporting in minutes:seconds and includes the 3 minute baseline, such that ignition was at 3:00 (clock time). Figure 10.4A shows a standard chair configuration 60 seconds after ignition. As the seating fire becomes established, it is then able to propagate to the arm rest assemblies, and quickly engulfs all of the soft combustible components in the seating area, as seen in Fig. 10.4B. However, the fire is restricted to just the seat region of the chair, with the lower half of the chair still unaffected by the fire. The foam and fabric from the seat back cushion that does not immediately burn will melt down onto the seat cushion where it starts to form a pool within the seat. For the more heat-resistant fabrics, the seat cushion, underside

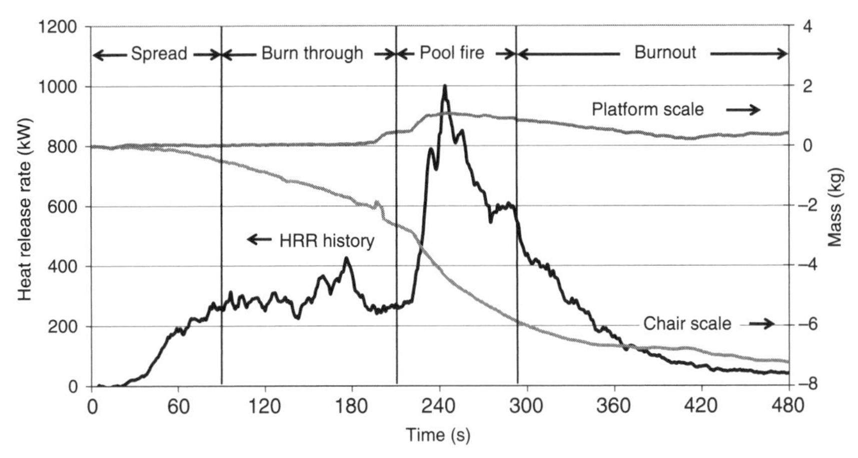

10.3 Mass and HRR histories for the wool fabric covering domestic furniture foam, showing the four general phases of burning: (i) Spread; (ii) Burn Through; (iii) Pool Fire, and (iv) Burnout.

fabric and sides, form a fabric pan that contains the molten material. In addition to the seat back foam and fabric, the seat cushion is also contributing to the 'seat pool fire'. All of the soft combustibles in the seating region that do not immediately burn, melt and accumulate on the seat cushion, where they pool before eventually burning. During the Spread phase there is little measurable mass of molten fuel on the platform below the chair, but the HRR tends to grow steadily as the fire spreads over the surface of the seating area.

Burn Through phase. This begins once the flame has spread over all of the surfaces including seat, back, and arm rests. Burn Through ends when the fabric on the underside of the seat cushion and/or webbing fails, spilling the molten foam and fabric onto the platform beneath the chair. The seat fire is present during this phase, as shown in Fig. 10.4C. As this phase proceeds, the seat cushion starts to show the initial signs of failing through the gradual increase in mass measured on the platform load cell, as seen in Fig. 10.3. During the Burn Through phase, the fire enters into a relatively steady burning region, as seen in Fig. 10.3 between 80 and 200 seconds. At 120 s, the ignition burner is switched off, sometimes causing a decline in the HRR. The length of time before the seat pool fire burns though the seat fabric and webbing is strongly dependent on the fabric material. Thermosetting fabrics that char, are supported by the webbing and are very effective at containing the molten foam fire in the seat region.

10.4 Standard CBUF Series 2 chair with wool fabric covering domestic furniture polyurethane foam. Photos (a) and (b) show the characteristics of the Spread Phase. Photo (c) shows the characteristics of the Burn Through Phase. Photos (d) and (e) show the Pool Fire Phase. Photo F shows the Burnout Phase.

Pool Fire phase. From the experimental observations, a combination of features has to occur before the Pool Fire phase can proceed. Firstly, the seat cushion and/or webbing must have failed, contributing significantly to the initial stages of the pool fire, and allowing the pool fire to become established below the chair. Secondly, some or all of the fabric on the front, back and side walls must have failed, allowing a plentiful supply of air to the pool fire. Figure 10.4D shows the transition from the Burn Through to

Pool Fire phases. The photo shown in Fig. 10.4D was taken just after the molten fuel spilled onto the platform below the chair. This is evident by close examination of the photograph; note that the remaining fabric on the front lower half of the chair is being backlit from the pool fire on the platform. Figure 10.4E shows the chair 225 seconds after ignition, as the fire is growing to the peak heat release rate. The rapid growth can be seen in both the HRR and platform mass curves in Fig. 10.3. As shown in Fig. 10.3, this phase is predominately short lived, in the order of seconds not minutes, and the HRR and platform load cells quickly climb to peak values before starting the Burnout phase.

Burnout phase. As the Burnout phase takes over, the HRR and platform load cell indicate the spilled molten mass starts to decline, once the bulk of the soft combustibles is consumed. Typically, there are no soft combustibles present on the burning frame, and any that are remaining are pooled on the platform below the furniture item. Figure 10.4F shows the chair 360 seconds after ignition, where only the remains of the fuel on the platform are burning. The Burnout phase can be seen in Fig. 10.3 as the characteristic gradual decay in the HRR curve and also the platform scale. This represents the depletion of the pool fire and the slow burning of the timber frame.

10.3.1 Alternative scenarios

The physical description of the burning phases given above is considered typical for furniture. However, for certain foam–fabric combinations, the description requires modification. The alternatives are the result of either readily combustible materials or highly resistant materials.

Readily combustible materials. For the more readily combustible combinations, i.e. thermoplastic fabrics covering non-combustion-modified foams, there may be the bypassing of the Burn Through phase as a distinct phase, because the chair quickly transitions into the Pool Fire phase. The flame quickly propagates around the seating area of the chair due to the poor thermal resistance of the fabric. The fabric rapidly melts and exposes the foam directly to the flame. The low ignition resistance of the materials means the seat pool fire quickly burns through the seat fabric and webbing, discharging the pool directly onto the platform. Once the pool fire is established on the platform, the fabric on the lower half of the chair is compromised, allowing free flow of air into the pool fire. The HRR for the chair shows a steep rise to peak values before the ignition burner is turned off. Figure 10.5 shows the heat release rate, chair mass, and platform mass histories for a standard chair with polypropylene fabric over domestic

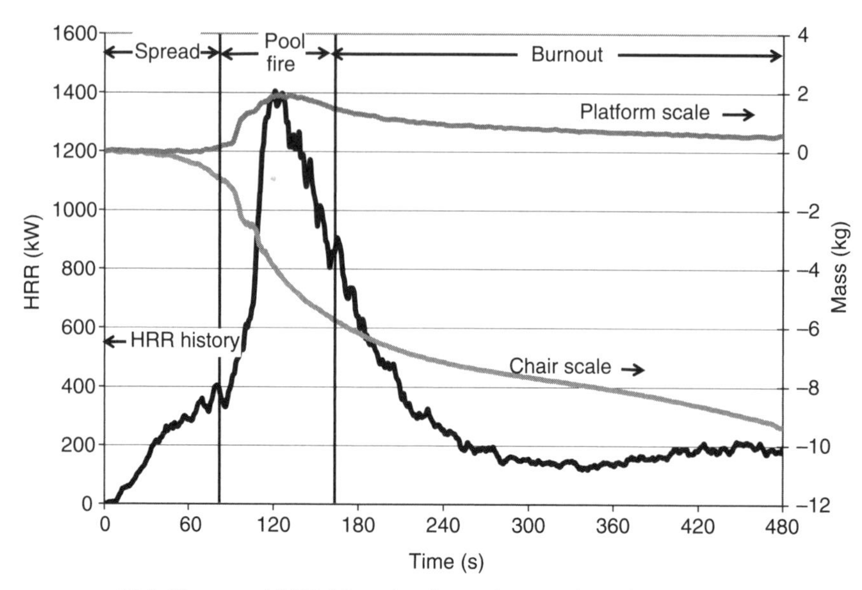

10.5 Mass and HRR histories for polypropylene fabric covering domestic polyurethane foam – showing Spread, Pool Fire, and Burnout. Burn Through is effectively nonexistent.

polyurethane foam, and includes the highlighted burning phases. For this case, there is no distinguishable Burn Through phase shown in Fig. 10.5. Figure 10.6 is a series of photographs taken during the experiment shown in Fig. 10.5. The photo in Fig. 10.6A, taken 45 seconds after ignition, shows the fire spreading over the chair and along the top of the arm rests during the Spread phase. The Spread phase in Fig. 10.5 shows almost continuous growth up to the Pool Fire phase. The fire developed so rapidly that the integrity of the seat failed quickly (before the burner was extinguished), spilling the contents of the seat onto the platform below. Once the integrity of the seat was compromised, the fire entered into the Pool Fire phase and quickly reached its peak heat release rate of 1400 kW at 120 s. Coincidentally, the burner gas flow was shut off around the same time. Figure 10.6B shows the fire 120 seconds after ignition during its peak heat release rate. The Pool Fire phase was short lived and the fire dropped into the Burnout phase as seen in the HRR plot in Fig. 10.5 and the photo taken 240 seconds after ignition in Fig. 10.6C.

Fire resistive materials. A chair with excellent fire resistive properties may not be able to support flaming once the gas flow to the burner is shut off and may self-extinguish during the Burn Through phase. This behaviour

(a) (b) (c)

10.6 Photographs showing different phases of burning for a polypropylene fabric covering domestic polyurethane foam. (a) – Spread Phase, (b) – Pool Fire Phase, (c) – Burnout Phase.

was observed for two out of three identical chairs with wool fabric covering (combustion modified) aviation foam. However, when a third wool fabric chair coupled with aviation foam was tested, the fire did eventually burn through and ignite the pool fire below. The chair reached its peak HRR at 594 s, after burning for more then 300 s with a heat release rate less than 5 kW.

Figure 10.7 shows the heat release rate and mass histories for a chair constructed with combustion modified foam covered with wool fabric along with the three burning phases observed in this experiment. In this case, the chair had the characteristic Spread, Burn Through, and Burnout phases. The spread phase lasted beyond 120 seconds when the gas flow to the burner was turned off. Once the gas flow to the burner was turned off, the HRR continued to decline as seen in Fig. 10.7. At approximately 150 seconds, a portion of the fabric on the underside of the seat cushion failed, spilling molten fuel onto the platform. However, the pool fire phase never eventuated as the molten material spilled onto the platform but never ignited. The chair self-extinguished around 360 s.

Figure 10.8 shows a series of photographs taken during each phase of the experiment presented in Fig. 10.7. Figure 10.8A, taken 45 seconds after ignition, shows the fire spread across the seat, up the back cushion, and onto the arm rest in a similar manner to the most combustible chairs; however, the development is noticeably slower. Figure 10.8B was taken 120 seconds after ignition, when the burner was shut off. It took another 30

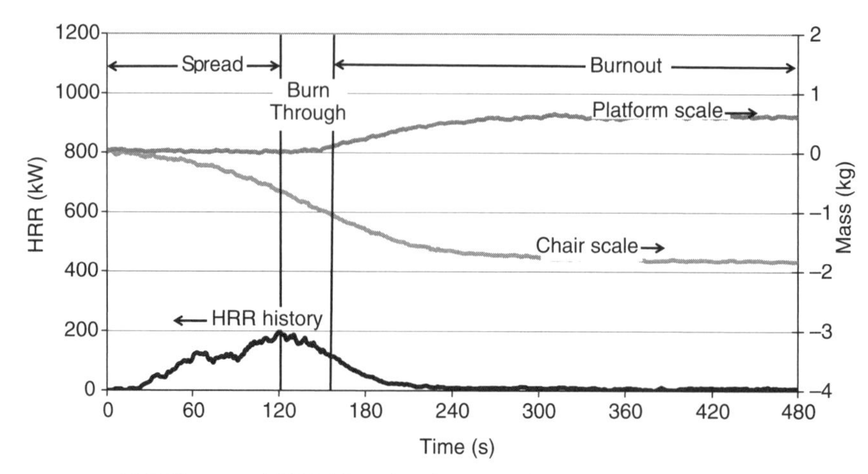

10.7 Mass and HRR histories for wool fabric covering combustion-modified foam showing Spread, Burn Through and Burnout. Pool Fire was effectively nonexistent because the pool did not ignite.

seconds before any measurable spilled mass was detected on the platform scale. Figure 10.8C shows the chair 180 seconds after ignition, when the fire was clearly into its Burnout phase. Figure 10.8D is the post fire photo, showing large portions of the seat and back foam still in place, although the wool fabric has burned through the fabric on the underside of the seat. The fabric on the back of the chair is still intact, although somewhat scorched.

10.4 Standard test methods for upholstered furniture and mattresses

There is a broad spectrum of standard test methods for upholstered furniture and mattresses covering a wide range of fire scenarios. Test standards can be broadly divided into three areas based on the item used in the test: component, mock-ups, or actual furniture. For example, some standards such as TB117 *Requirements, Test Procedures and Apparatus for Testing the Flame Retardance of Resilient Filling Materials Used in Upholstered Furniture*[6] use only individual components, namely the padding material and fabric. The test is a small, bench-scale test that exposes the sample to a small laboratory flame. In this test, individual components are exposed to the burner flame, but there is no radiant feedback as in a real fire. Therefore, this test is not representative of a real fire. A separate procedure is also included in the standard for exposing the padding material to a burning cigarette. This style of test has proved to be a poor predictor of the fire performance of the finished furniture item and is currently under

10.8 Series of photos showing a wool fabric covering the combustion-modified aviation foam. (a) – Spread Phase, (b) – Burn Through Phase, (c) – Burnout Phase, (d) – Post fire, showing large portions of the seat and back foam still in place.

review. Other standards, such as BS 5852:1990 *Methods of Test for Assessment of Ignitability of Upholstered Seating by Smoldering and Flaming Ignition Source*[7] rely on mock-ups constructed using the actual covering and padding materials together, and use simple pass/fail criteria based on ignition of the mock-up. More detail about BS5852 is given below. The most applicable test standards involve testing of actual furniture items and quantifying the actual fire hazard using Oxygen Depletion Calorimetry (ODC). TB133 *Flammability Test Procedure for Seating Furniture for Use in Public Occupancies*[4] is an example of a regulatory standard used in

California, USA, which specifies the heat release rate as part of the pass/fail criteria. The California Bureau of Home Furnishing and Thermal Insulation (CBHFTI) has been in the vanguard in terms of developing regulations for controlling the flammability of upholstered furniture and mattresses.

For more than a decade, the Consumer Product Safety Commission (CPSC) has been involved in a major research effort to determine the most appropriate way to reduce fire deaths from upholstered furniture and mattress fires on a national level within the USA. On July 1, 2007 a new open flame mattress standard (16CFR § 1633) went into effect, requiring all manufactured, imported or renovated mattresses to meet this standard. Recently, the CPSC voted to issue a Notice of Proposed Rulemaking (NPR) on a mandatory standard for controlling the flammability of residential upholstered furniture.

Table 10.1 gives a summary of many of the standard test methods used for regulating upholstered furniture. The table summarizes the sample type, ignition sources, and pass/fail criteria. In several cases, identical or slightly modified versions of the standards are available from other standards writing authorities. Table 10.1 is intended to give an overview of some of the standards and not a comprehensive comparison. Specific details about the standards can be obtained from the actual documents, which are given in references. A summary of some of the more relevant standards are given below.

10.4.1 Standard tests for assessing upholstered furniture flammability

BS 5852:1990 Methods of test for assessment of the ignitability of upholstered seating by smoldering and flaming ignition sources[7]

The objective of the standard is to measure the ignitability of upholstered furniture composites and/or complete pieces of furniture. The standard incorporates eight ignition sources including a smoldering cigarette, three butane flames, and four burning wooden cribs. The strength of the flaming sources of ignition are designed to approximate in strength between a lit match and the burning of four double sheets of a full-size newspaper. The standard allows for the composites of the furniture materials to be tested rather then the entire piece of furniture.

Figure 10.9 shows the geometry defined in the standard. For the cigarette and the small open flame test, the horizontal cushion is 150 mm by 450 mm and the vertical cushion is 300 mm high and 450 mm wide. For the wooden crib experiments, the horizontal cushion is 300 mm by 450 mm and the vertical is 450 mm high and 450 mm wide.

Table 10.1 Summary of upholstered furniture and mattress standard test methods comparing the type of sample, ignition sources used and acceptance criteria

Test method	Sample type			Ignition source			Acceptance criteria						
	Component	Mock-up	Finished product	Cigarette	Small flame	Large flame	Ignition	Progressive smoldering	Char length	Flame spread	Heat release rate	Total heat release	Other
Upholstered furniture tests													
BS 5852		X	X	X	X	X	X	X		X			
TB 116			X	X			X	X	X				
TB 117	X				X				X	X			
TB 133			X			X					X	X	X[3]
Mattress tests													
BS 6807		X	X	X	X	X	X	X		X			
TB 106			X	X			X	X	X				
TB 603			X			X					X	X[4]	
16CFR § 1633													
TB 121[1]			X			X							X[5]
TB129[2]			X			X					X	X	X[6]

[1]Technical Bulletin 121, *Flammability Test Procedure for Mattresses for Use in High Risk Occupancies*, California Bureau of Home Furnishings and Thermal Insulation, North Highlands, CA, USA, 1980.
[2]Technical Bulletin 129, *Flammability Test Procedure for Mattresses for Use in Public Buildings*, California Bureau of Home Furnishings and Thermal Insulation, North Highlands, CA, USA, 1992.
[3]Pass/Fail criteria include ceiling temperature, temperature 1.2 m above floor, smoke opacity, CO concentration and mass loss.
[4]Although the test procedures are nearly identical, acceptance criteria for the total heat release in the first 10 minutes is 15 MJ in 1633 versus the less strict TB603 limit of 25 MJ.
[5]Test criterion also includes total mass loss, temperature above mattress, and CO concentration in the compartment.
[6]Pass/Fail criteria also include mass loss.

10.9 Photograph showing BS5852 upholstered furniture mockup during experiments; ignition from small butane flame.

For the smoldering ignition, the lit cigarette is placed at the junction between the vertical and horizontal cushions. The test lasts for 60 minutes, during which there can be no ignition or progressive smoldering observed. If there is no observable ignition or progressive smoldering, then the test is repeated at another location on the same sample. For the small butane flame test, the flame height and exposure time ranges from approximately 35 mm for 20 seconds (Source 1), 145 mm for 40 seconds (Source 2) to 240 mm for 70 seconds (Source 3). The flame is placed at the junction between the vertical and horizontal cushions. The same pass/fail criteria apply, i.e. no observable ignition or progressive smoldering. If the sample fails to ignite or progressively smolder, then the test is repeated. If the sample fails to ignite or progressively smolder in the repeated test, then the sample is considered to have 'no ignition'. For the wooden cribs, (ignition Source 4–7), a progressively larger wooden crib is placed on a new sample for each test. Source 4 (the smallest crib) is 40 mm square assembled with 5 stacked layers of 6.5 mm sticks. Source 5 is identical to Source 4 but has 10 layers of sticks. Source 6 is 80 mm square and assembled from 4 layers of 12.5 mm sticks. Source 7, the largest crib, is identical to Source 6 but is nine layers high. The crib is placed on the horizontal cushion and against the vertical cushion. The crib tests are 60 minutes long, during which there can be no observed flaming 10 minutes after the crib ignition for Sources 4 and 5 or 13 minutes after ignition for the larger cribs used for Sources 6 and 7. BS5852 has been replaced with BSEN 1021 which is an adaptation of ISO 8191,[8] but the basic principles remain the same.

TB133 Flammability test procedure for seating furniture for use in public occupancies[4]

The objective of TB 133 is to test seating for use in occupancies that are identified for public use, such as prisons, health care facilities, public auditoriums, and hotels. It is not intended for residential use. The test is conducted in a 3 m × 3.6 m × 3 m high compartment. The only ventilation is a 0.95 m wide by 2 m high door. The test can be conducted in a smaller compartment such as ISO9705[9] room, when equivalent test results can be demonstrated. A full scale furniture item is placed in the corner opposite the door. Ignition is by a 250 m square tube burner generating approximately 18 kW for 80 seconds. For furniture items 1 meter or less, the burner is centered on the seat area, 25 mm above the seat surface, 50 mm from the rear vertical cushion. For items larger than 1 meter, the tube burner is placed 125 mm from the left arm of the item. The test is allowed to continue until all combustion has ceased, 1 hour has elapsed, or flameover or flashover appears to be inevitable. The pass/fail criterion allows for either measurements within the room or from oxygen depletion calorimetry (ODC). The criterion varies slightly, depending on the measurements used. The criteria are summarized here; however, further detail on the pass/fail criteria can be found in the actual standard.[4] The furniture item fails if any of the following criteria are exceeded within the room:

- temperature increase of 111 °C or greater at ceiling thermocouple
- temperature rise of 28 °C or greater at the 1.2 m thermocouple
- opacity greater than 75% at the 1.2 m-high smoke opacity meter
- CO concentration in the room of 1000 ppm or greater for 5 minutes
- weight loss due to combustion of 1.36 kg or greater in the first 10 minutes

The furniture item fails if any of the following criteria are exceeded within the room test using ODC:

- maximum heat release rate of 80 kW or greater
- total heat release rate of 25 MJ or greater in the first 10 minutes
- greater than 75% opacity at the 1.2 m-high smoke opacity meter
- CO concentration in the room of 1000 ppm or greater for 5 minutes

10.4.2 Standard tests for assessing mattresses

CAL TB 106 Requirements, test procedures and apparatus for testing the resistance of a mattress or mattress pad to combustion which may result from a smoldering cigarette (Federal Standard 16 CFR 1632)[10]

The objective of TB 106 is to assess the resistance to ignition of a mattress from a burning cigarette. The test procedure requires a finished product

10.10 Burner location for CAL TB603 test standard for mattress/foundation exposed to open flame.

ready for sale to the consumer. The procedure requires that three burning cigarettes be placed on the smooth surface of the mattress, three along the tape edge, three on quilted locations and three on tufted locations. Tests are carried out with the burning cigarettes on the bare mattress and sandwiched between two cotton sheets over the mattress. The mattress fails the test if any of the following conditions occur:

- obvious flaming combustion
- smoldering combustion continuing for more than 5 minutes after the cigarette has extinguished
- char developing more than 50 mm in any direction from the cigarette

CAL TB 603 Requirements and test procedure for resistance of a mattress/box spring set to a large open-flame (Federal Standard 16 CFR 1633)[11,12]

The objective of this standard is to determine the burning behaviour of a mattress plus foundation. Derived from extensive research conducted by the National Institute of Standards and Technology (NIST), the mattress, along with the foundation, are placed on top of a bed frame that sits over a catch surface. The mattress is exposed to two gas burners, one on top and the other on the side, as seen in Fig. 10.10. The burners are designed to mimic the local heat flux imposed on a mattress/foundation from burning bedclothes. The top burner flows 19 kW for 70 s and the side burner flows 10 kW for 50 s. The test can be conducted in one of three configurations: Open Calorimeter; 2.4 m × 3.6 m × 2.4 m high room; or 3.05 m × 3.66 m

× 2.44 m high room. A large extraction system with oxygen depletion calorimetery is required. The test can last 30 minutes unless all signs of burning have ceased or the fire develops to such a size that it must be extinguished. The mattress is considered to have failed if the peak heat release rate exceeds 200 kW or the total heat released exceeds 25 MJ in the first 10 minutes.

BS 6807 Assessment of the ignitability of mattresses, divans, and bed bases with primary and secondary sources of ignition[13]

The objective of BS6807 is to assess the ignitability of a mattress. The test procedure uses the same ignition sources as specified in BS 5852: a cigarette, small gas flames and wood cribs. The standard specifies procedures for primary ignition of the mattress, i.e. the ignition source placed directly onto the surface of the mattress; and for secondary ignition the ignition source is placed onto the bedcovers. Bedcovers include sheets, blankets, bedspreads, valances, quilts and mattress covers.

For the smoldering ignition, the lit cigarette is placed on top of the mattress. For the flaming sources the standard requires testing with the source on top of the mattress and underneath the mattress. The precise location of the ignition source is mapped out in detail within the standard. The same pass/fail criteria used in BS 5852 apply, i.e. no observable ignition or progressive smoldering as defined in the standard.

10.5 Future test methods

On March 4, 2008 the Consumer Product Safety Commission released the Notice of Proposed Rulemaking (NPR) Proposed Rules 16 CFR Part 1634 *Standard for the flammability of residential upholstered furniture*.[14] The proposed rules can be downloaded at http://www.cpsc.gov/businfo/frnotices/fr08/furnflamm.pdf. The CPSC activity resulted from a petition to the CPSC from the National State Fire Marshals and a summary of the background can be found in the NPR. In brief, the standard is a smoldering ignition test on the covering material and does not address the flammability of the filling material. The proposed standard provides for two approaches: (1) a covering material that complies with the prescribed smoldering ignition resistance test or (2) an interior barrier that complies with a specified smoldering and small open flame ignition resistance test.

The proposed smoldering test method for *Type 1* (furniture with complying fabric cover) places the covering fabric over standard polyurethane foam (SPUF) panels. The test requires two panels, vertical and horizontal, each covered with the fabric to be tested and placed in the specimen holder with a lit cigarette placed in the crevice between the horizontal and

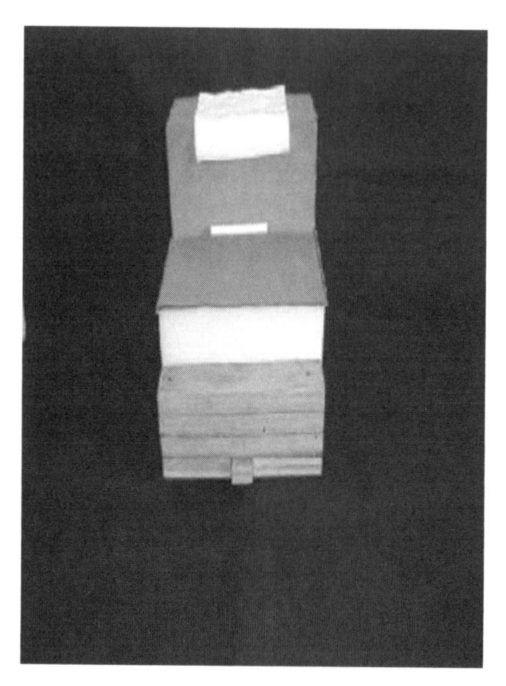

10.11 Smoldering test specimen holder for CPSC NPR 16 CFR Part 1634 Standard for the Flammability of Residential Upholstered Furniture; proposed rule.

vertical panels. Figure 10.11 shows the cover fabric over the SPUF in the specimen holder with the cigarette in place. The cigarette is then covered with a piece of sheeting fabric and allowed to burn over its entire length. The test lasts 45 minutes, with observations and measurements recorded. The specimen must not continue to smolder after the test nor transit to flaming. The SPUF must not lose more than 10% of its mass and replicate samples are to be tested. If any of the initial specimens fail, an additional 20 specimens are to be tested and at least 25 of the 30 specimens tested must meet the criteria.

If the cover fabric does not meet the smoldering test described above, a complying interior barrier can be used under the fabric – *Type 2* furniture. A complying barrier must not only meet the criteria for the smoldering test but also an additional small open flame test. For the barrier smoldering test, the interior barrier is placed between the SPUF and a standard cover fabric: velvet pile, 100% cotton, beige in colour.

The open flame test setup is similar to that of BS5852, in which a mock-up is constructed with the barrier fabric placed over the SPUF and covered with a standard fabric made from 100% rayon with a 20/2 ring spun basket weave. The mock-up is placed on a metal frame and a 240 mm butane flame is placed in the crevice of the mock-up for 70 seconds. The sample is observed for 45 minutes and the sample must not lose more than 20% of the mass of the mock-up. Like the smoldering test, initially ten replicate tests are conducted and all ten mockups must meet the acceptance criteria. If any of the initial specimens fail, an additional 20 specimens are to be tested and at least 25 of the 30 specimens tested must meet the criteria.

At the time of writing, it was not clear if this proposed standard will be adopted by the CPSC. There has been significant opposition to the standard from the California Bureau of Home Furnishings, Thermal Insulation, the National Association of State Fire Marshals and several other interested parties. The public comments on the proposed standard can be found on the CPSC website: http://www.cpsc.gov/LIBRARY/FOIA/FOIA08/pubcom/flamm1.pdf.

10.6 Strategies for reducing the fire threat of upholstered furniture

There are a number of strategies that can be employed to reduce the fire threat from upholstered furniture. Common strategies can be divided into two methodologies: use slow burning materials (padding and/or fabric) or protect the padding material from the ignition source.[15] The recent regulation and research on upholstered furniture has prompted development of new and improved materials for use in furniture. The adoption of CFR Part 1634 *Standard for the flammability (open flame) of mattress set*[12] has contributed to the development of new barriers and improved manufacturing techniques. However, in many cases the improvements in manufacturing are often considered proprietary information. Fortunately results of the European Commission study *Combustion Behaviour of Upholstered Furniture* (CBUF)[24] have been published and present insights into improving furniture design. These results are summarized briefly here and the reader is directed to this work for more information.

10.6.1 Material selection

Fabric

Fabric can be broadly divided into two categories, those that melt and those that char. Melting fabrics, made from thermoplastic materials such as polyester, nylon, polypropylene, polyacrylic, melt away from the ignition source and expose the padding material, which typically leads to rapid fire growth as described in Section 10.3. Charring fabrics, made from natural materials such as cotton, wool, viscose, will form a char layer that can provide some level of protection to the padding material from the ignition source and prevent, or at least delay, the involvement of the padding material. However, charring materials exhibit a greater tendency to smoldering combustion and can be more susceptible to smoldering ignition from cigarettes. Many of today's fabrics are blends and tend to exhibit some characteristics of both charring and melting fabrics. There are also a number of fabrics that have been treated either by back coating or by using an additive with fire retardant properties to improve the fire performance. Fire retardant treatments require careful consideration due to their environmental and health concerns. Typically, the selection of a fabric is an aesthetic choice rather than a fire safety consideration, but with careful consideration the threat posed by upholstered furniture can be reduced.

Filling

The vast majority of furniture uses foam as the filling material. There are many different types of foam with the primary type being polyurethane. Foam is categorized by density, hardness and combustion properties. In a simple sense, reducing the density of the foam will reduce the fire hazard by reducing the amount of fuel available to burn. However, this may not give the hardness or wear resistance desired. The use of combustion-modified foam has shown significant improvement in the fire performance, especially when coupled with a natural fabric such as wool. Reducing the rate of energy release from the foam can give a very positive reduction in fire hazard for upholstered furniture, especially for open flame ignition sources.

Interliners

Interliners, intermediate layers between the cover fabric and the foam, are common in furniture; polyester wadding is often used to give a desired cushion appearance compared to the blocky look of foam. Polyester wadding has only a small effect on the fire performance of upholstered furniture. Commercial airlines have been using interliners for many years to improve the fire performance of aircraft seating. Interliners are also now being routinely used in mattresses in the US since CPS standard 16 CFR Part 1633 *Standard for the flammability (open flame) of mattress sets; final rule*[12] came into effect. The impact of interliners is most pronounced for open flame ignition sources.[16]

Frames

Although there are many different types of frame found in upholstered furniture, wood is by far the most common. Typically in furniture fires, the wood makes only a minor contribution to the peak heat release rate or fire growth rate. The exception to this is when the wood frames are constructed with corrugated metal plates that can fail early in the fire and expose large virgin upholstery suddenly to the fire, resulting in rapid fire growth.[24] Furniture that uses thermoplastic frames have not been well studied but research indicates that thermoplastic chairs do pose a greater fire hazard.[17] Chairs without frames, such as bean bags and foam blocks chairs, can be very hazardous due to large amounts of unrestrained easily combustible material.

10.7 Modelling of furniture fires

The ultimate goal of furniture/bedding fire modelling is to be able to predict the fire behaviour of the furnishings with a minimal amount of experimentally derived data. In an ideal world the user could simply go to a database and download the necessary properties for a desired furniture item and run their model to predict the fire growth rate, perform a forensic analysis, or carry out a hazard assessment based on the results from a Computational Fluid Mechanics model (CFD) or zone model. For details on compartment fire modelling see references 18–20.

Unfortunately, the modelling of furniture and bedding fires is still in its infancy. The burning behaviour of upholstered furniture and bedding creates significant challenges for fire modellers beyond simply the solid phase burning characteristics. Upholstered furniture and mattresses are often complex composites, constructed from at least three materials – padding, covering and framing. The problem is further complicated when most commonly used soft materials and common construction techniques lead to complex burning phenomena, including burn through, tunnelling, melting, flowing, and forming pool fires remote from their original locations. Modelling efforts have been developed using experimental correlations[21] and physics-based models to predict the burning behaviour of upholstered furniture.[22,23] In these models, the researchers have had to over-simplify the problem, leaving out many of the important details of the burning item. However, useful information can still be obtained from these models and also simple correlations that can be utilized in developing less combustible furniture.

10.7.1 Combustion behaviour of upholstered furniture (CBUF)[24]

The CBUF research program was conducted by a consortium of research organizations, one each from Belgium, Denmark, Finland, France, Germany, Italy, Sweden, and three from the UK. The program is the most extensive study on upholstered furniture and bedding ever conducted. There were 71 room tests, 154 furniture calorimeter tests, and 1098 composites tested in the cone calorimeter. This does not include all of the preliminary experiments to establish the ignitability, effects of room size, ventilation, test procedures, repeatability and reproducibility. Details about the CBUF program can be found in the comprehensive final report given in reference 24. The primary findings of the research are summarized below:

- The Heat Release Rate from the full-scale calorimeter experiments on full-scale furniture items was identified as the primary measure of the fire hazard from upholstered furniture. The test procedure for the furniture calorimeter was substantially similar to that of TB133. The TB133 tube burner was used as the ignition source; however, the HRR of the burner was increased to 30 kW for 120 s. This thermal insult to the furniture item assured ignition of most items.
- Three models were developed for predicting the HRR from an item of upholstered furniture burned in the furniture calorimeter using the results from foam–fabric composites in the cone calorimeter.
  (i) Model I is a set of correlations developed to predict the peak HRR, time to peak, and time to untenable conditions in an ISO9705 room from an item of upholstered furniture burned in the furniture calorimeter. The correlations use the results from foam–fabric composites in the cone calorimeter. This correlation can help to reduce the number of full-scale tests required.
  (ii) Model II is a combined physics/correlation model that uses an area-deconvolution model (based on an earlier flame spread model by Wickstrom and Goransson[25]) for predicting full scale HRR time histories from results of the foam–fabric composite cone calorimeter tests and representative furniture calorimeter test.
  (iii) Model III is a physics-based model developed for predicting the HRR of burning mattresses only.
- Existing room fire models can be used to predict the escape time and interface height using the full scale HRR results. However, it is up to the regulator to determine the allowable escape time from the room of fire origin.

The CBUF Model I (described in detail in references 24 and 26) is a factor-based method that uses a series of statistically correlated factors to predict the peak HRR, total heat release, time to peak, and time to untenability. The model is an improvement on the earlier (1985) factor-based prediction from NIST.[27] The original model was examined for applicability to the CBUF items. It was found to apply only generally and tended to

Table 10.2 Furniture style factors defined in the CBUF predictive model

Code	Style factor A	Style factor B	Type of furniture
1	1.0	1.0	Armchair, fully upholstered, average amount of padding
2	1.0	0.8	Sofa, two-seat
3	0.8	0.9	Sofa, three-seat
4	0.9	0.9	Armchair, fully upholstered, high amount of padding
5	1.2	0.8	Armchair, small amount of padding
6	1.0	2.50	Wingback chair
7	–	–	Office swivel chair, plastic arms (unpadded), plastic back shell
8	–	–	High-back office swivel chair, plastic arms (unpadded), plastic back, and bottom shell
9	–	–	Mattress, without inner spring
10	–	–	Mattress with inner spring
11	–	–	Mattress and box spring (divan base) set
12	0.6	0.75	Sofa-bed (convertible)
13	1.0	0.80	Armchair, fully upholstered, metal frame
14	1.0	0.75	Armless chair, seat and back cushions only
15	1.0	1.00	Two-seater, armless, seat and back cushions

under-predict the more modern and varied European furniture. The study undertook further development and refinement of this model. A series of differing furniture styles constructed from the same 'soft' combustible material combinations (soft being the foam, fabric, and interliner) were tested. An analysis of the results brought about several refinements from the 1985 NIST model to the CBUF Model I. Notably, the mass of soft combustibles replaced the mass of total combustibles, and the power was raised from 1 to 1.25. The time to ignition in the cone calorimeter test was seen as an important variable and was included.

The style factor also required significant change to account for the new European furniture. Incorporated in the calculation of the peak heat release rate and time to peak, the style factor accounts for the physical differences that cannot be resolved by the cone calorimeter test method, including the ornate and intricate detail that can be found in some furniture. Table 10.2 provides the style factors needed in the predictive model.

Peak HRR

Equation [1] is the first correlating variable for the peak heat release rate. It was found that the partially correlating variable x_1 represented well the general trend with the exception of groupings of high peak HRR (over 1200 kW). Considering only these data points, the second correlating variable x_2 emerged as given in Equation [2]. The symbols along with their units are defined in Section 10.9, Nomenclature.

$$x_1 = (m_{soft})^{1.25} \cdot (\text{style\_fac.A})(\dot{q}''_{pk} + \dot{q}''_{300})^{0.7} (15 + t_{ig})^{-0.7} \quad [1]$$

$$x_2 = 880 + 500 \cdot (m_{soft})^{0.7} (\text{style\_fac.A}) \cdot \left(\frac{\Delta h_{c,\,eff}}{q''} \right)^{1.4} \quad [2]$$

The input data used in Equations [1] and [2] are from the cone calorimeter test conducted in accordance with the strict CBUF protocol. Selection rules are established that are termed 'regimes', to determine when to use x_1 and x_2, with x_1 displaying a partial dependence.

Regimes:

{1} If (x_1 >115) or ($q'' > 70$ and $x_1 > 40$) or (style = {3,4} and $x_1 > 70$) then,

$$\dot{Q}_{peak} = x_2 \quad [3]$$

{2} else If, $x_1 > 56$ then,

$$\dot{Q}_{peak} = 14.4 \cdot x_1 \quad [4]$$

{3} else,

$$\dot{Q}_{peak} = 600 + 3.77 \cdot x_1 \quad [5]$$

Total heat release

The total heat release is determined from the actual mass of the furniture item and small-scale effective heat of combustion. Differentiation is noted between the 'soft' and total combustible masses. Experimental observation reveals that the affect of a wooden frame is not seen until nearly all of the 'soft' materials are consumed. Equation [6] was found to represent the total heat release:

$$Q = 0.9 m_{soft} \cdot \Delta h_{c,\,eff} + 2.1 (m_{comb,\,total} - m_{soft})^{1.5} \quad [6]$$

Time to peak

The time to peak heat release is as important as the peak heat release rate in hazard calculations. Equation [7] was developed to predict time to peak

HRR from sustained burning (50 kW). It is recognized that often other hazard variables are maximized at or near the time of peak HRR.

$$t_{pk} = 30 + 4900 \cdot (\text{style\_fac.B}) \cdot (m_{soft})^{0.3} \cdot (\dot{q}''_{pk\#2})^{-0.5} \cdot (\dot{q}''_{trough})^{-0.5} \cdot (t_{pk\#1} + 200)^{0.2} \quad [7]$$

Note that a different style factor is incorporated into the time to peak calculation.

Time to untenable conditions (untenability time)

Equation [8] was developed to predict time to untenable conditions in a standard room. Untenability time is defined as the time from 50 kW HRR to 100 °C temperature, 1.1 to 1.2 m above floor level.

$$t_{UT} = 1.5 \times 10^5 (\text{style\_fac.B})(m_{soft})^{-0.6} (\dot{q}''_{trough})^{-0.8} (\dot{q}''_{pk\#2})^{-0.5} (t_{pk\#1} - 10)^{0.15} \quad [8]$$

A more complete discussion of the model and a discussion of the accuracy can be found in the original CBUF report.[24] A subsequent study in New Zealand[29] on eight different furniture items did not show as encouraging results as the original work.

10.7.2 Intrinsically safe upholstered furniture

When assessing the hazard from flaming combustion of furniture, the most important consideration is whether or not the fire will propagate over the surface of the item. For a propagating item, the fire will spread over the surfaces and consume most of the soft combustibles, although often large pieces of the frame may survive. The propagating behaviour can be seen in Fig. 10.6. For nonpropagating items, the fire behaviour is quite different. Once the ignition source is removed or burns out, the fire is not self-sustaining and will burn out, leaving most of the soft combustibles unburned. Based on the experimental observations in the CBUF study, non-propagating full-scale chair fires typically have a heat release rate of between 20 and 100 kW. This is consistent with the pass/fail result of 80kW used in TB133. Comparing the results from the cone calorimeter experiments in the rigorous CBUF cone calorimeter protocol, it was found that a value of $\dot{q}''_{180} = 65$ kW/m^2 or less correlated well with the nonpropagating items.

The total HRR and emission of toxic products for a nonpropagating item are expected to give substantially similar results to those measured in the furniture calorimeter. This is based on experimental results which demonstrate that for a nonpropagating furniture item (i.e. produces less than

100 kW), the heat and radiation levels are too low to cause serious injury to occupants of the room unless they are intimate with the fire. It can therefore be concluded that nonpropagating furniture items are as *intrinsically safe*[1] as a combustible item can be.

However, for propagating items, the results from the furniture calorimeter for the total HRR and toxic product emission are considered to be lower limits of what can be expected in a real fire. This is due to the fact that a propagating item is capable of spreading to an adjacent combustible item that was not present in the test. In addition, independent analysis of the CBUF room experimental results showed that room interaction effects can be found when the item's HRR is as low as 500 kW[5]. Room interaction can result in an increased HRR from the radiation feedback from the room and can increase the production of toxic species by reducing the available oxygen, thus increasing the CO production.

10.8 Sources of further information and advice

This chapter has served as only an introduction to a very complex topic, with a large body of research having already been conducted but also an even greater amount of research still required. There are some excellent books available on upholstered furniture and mattress fires. Most notable is *Fire Behavior of Upholstered Furniture and Mattresses*[5] by Krasney *et al.*, which includes over 500 references. For a more general reference on heat release rate from burning objects, which includes several chapters on upholstered furniture and mattresses is *Heat Release in Fires*,[28] edited by Babrauskas and Grayson. The original CBUF Final Report[24] and subsequent research can be found in References 5, 29, and 30.

Some useful websites on the flammability of upholstered furniture and mattresses are:

Combustion Behaviour of Upholstered Furniture (CBUF) http://www.sp.se/EN/INDEX/RESEARCH/CBUF

Building Fire Research Laboratory – National Institute of Standards and Technology – includes copies of research reports and videos from a wide range of fire research including upholstered furniture and mattresses. http://www.bfrl.nist.gov/

California Bureau of Furnishings and Thermal Insulation – includes copies of technical bulletins on testing furniture and mattresses. http://www.bhfti.ca.gov

Consumer Product Safety Commission – contains up-to-date information on developments in possible regulations on the flammability testing of furniture and mattresses. http://www.cpsc.gov/

Upholstered Furniture Action Council – furniture manufacturers association includes several voluntary standards for furniture flammability. http://www.homefurnish.com/UFAC/index.htm

10.9 Nomenclature

$\Delta h_{c,eff}$	effective heat of combustion of the bench-scale composite sample (MJ/kg)
m_{soft}	mass of the soft combustible material of the full-scale item (kg)
$m_{comb,total}$	mass of the total combustible material of the full-scale item (kg)
q''	total heat release per unit area of the bench-scale composite sample (MJ/m^2)
$\dot{Q}_{peak}$	peak HRR, measured or predicted, of the full-scale item (kW)
Q	total heat release rate of the full-scale item (MJ)
$\dot{q}''_{pk}$	peak HRR per unit area of the bench-scale composite sample (kW/m^2)
$\dot{q}''_{pk\#2}$	second peak HRR per unit area of the bench-scale composite sample (kW/m^2)
$\dot{q}''_{trough}$	trough between two peak HRR, per unit area of the bench-scale composite sample (kW/m^2)
$\dot{q}''_{300}$	HRR per unit area (bench scale) averaged over 300 s from ignition (kW/m^2)
t_{ig}	time to ignition of the bench scale composite sample (s)
t_{pk}	time to peak in the full-scale item (s)
$t_{pk\#1}$	time to characteristic 'first' peak of the bench scale composite sample (s)
t_{UT}	time to untenable conditions in a standard room (s)

10.10 References

1. BABRAUSKAS, V., 'Upholstered Furniture and Mattresses', *Fire Protection Handbook*, 18th Ed, Section 8 Chapter 17, National Fire Protection Association, Quincy, Massachusetts, 2002.
2. CHOWDHURY, R., GREENE, M. and MILLER, D., *2003–2005 Residential Fire Loss Estimates*, Consumer Product Safety Commission, Aug 2008.
3. FLEISCHMANN, C. M. and HILL, G. R., *Burning Behaviour of Upholstered Furniture,* Interflam '04, Interscience Communications Co., London, 2004.
4 TECHNICAL BULLETIN 133, *Flammability Test Procedure for Seating Furniture in for Use in Public Occupancies*, California Bureau of Home Furnishings and Thermal Insulation, North Highlands, CA, USA, 1991.
5. KRASNY, J. F., PARKER, W. J. and BABRAUSKAS, V., *Fire Behavior of Upholstered Furniture and Mattresses*, Noyes Publications, New Jersey, USA 2001.
6. TECHNICAL BULLETIN 117 *Requirements, Test Procedures and Apparatus for Testing the Flame Retardance of Resilient Filling Materials Used in*

Upholstered Furniture, California Bureau of Home Furnishings and Thermal Insulation, North Highlands, CA, USA, 2000.
7. BS 5852:1990 *Methods of test for assessment of the ignitability of upholstered seating by smouldering and flaming ignition sources*, British Standard Institute, 1990.
8. ISO 8191 *Furniture Assessment of the Ignitability of Furniture, Part 1, Ignition Sources: Smouldering Cigarette, Part 2: Igntion Source: Match Equivalent Flame*, International Standards Organization, Geneva.
9. ISO 9705 *Fire Tests – Full-scale Room Test For Surface Products*, International Standard Organization, Geneva, 1993.
10. TECHNICAL BULLETIN 106 (16 CFR 1632) *Requirements, Test Procedures and Apparatus for Testing the Resistance of a Mattress or Mattress Pad to Combustion which may Result from a Smoldering Cigarette*, California Bureau of Home Furnishings and Thermal Insulation, North Highlands, CA, USA, 1986.
11. TECHNICAL BULLETIN 603 *Requirements and Test Procedure For Resistance of a Mattress/Box Spring Set to a Large Open-flame*, California Bureau of Home Furnishings and Thermal Insulation, North Highlands, CA, USA, 2004.
12. 16 CFR Part 1633 *Standard for the Flammability (Open Flame) of Mattress Sets; Final Rule*, Consumer Product Safety Commission, 15 March 2006.
13. BS 6807:1990, *Assessment of the ignitability of mattresses, divans, and bed bases with primary and secondary sources of ignition*, British Standard Institute, 1990.
14. 16 CFR Part 1634 *Standard for the Flammability of Residential Upholstered Furniture*, Consumer Product Safety Commission, 4 March 2008.
15. ANDERSSON, B. and MAGNUSSON, S. E., 'Fire behaviour of upholstered furniture – an experimental study', *Fire and Materials*, **9**, 41–45, 1985.
16. EGGESTAD, J., 'Effects of Interliners on the Ignitability of Upholstered Furniture', *Journal of Fire Sciences*, **5**, 152–161, 1987.
17. BABRAUSKAS, V. and KRASNY, J. F., *Fire Behaviour of Upholstered Furniture* (NBS Monograph 173), U.S. National Bureau of Standards, 1985.
18. JANSSENS, M. L., *An Introduction to Mathematical Fire Modelling,* Technomic Publishing, APA, USA, 2000.
19. *Fire Dynamics Simulator (Version 4)* Technical Reference Guide, NIST Special Publication 1018, U.S. Government Printing Office, Gaithersburg, USA, 2004.
20. KARLSSON, B. and QUINTIERE, J. G., *Enclosure Fire Dynamics*, CRC Press LLC, London, 2000.
21. BABRAUSKAS, V., MYLLYMÄKI, J. and BAROUDI, D., 'Predicting Full Scale Furniture Burning Behavior from Small Scale Data', Ch 8, *CBUF: Fire safety of upholstered furniture – the final report on the CBUF research program.* Director-General Science, Research and Development (Measurements and Testing). European Commission. *Report EUR 16477 EN*, 1995.
22. DIETENBERGER, M. A., *Technical Reference and User's Guide for FAST/FFM (Version 3)*, NIST-GCR-91-589, NIST, Gaithersburg, MD, 1991.
23. PEHRSON, R., *Prediction of Fire Growth of Furniture Using CFD*, PhD Thesis, Worcester Polytechnic Institute, Worcester, MA, USA.
24. SUNDSTROM, B. (Ed.) CBUF: Fire safety of upholstered furniture – the final report on the CBUF research program. Director-General Science, Research

and Development (Measurements and Testing). European Commission. *Report EUR 16477 EN*, 1995.
25. WICKSTROM, U. and GORANSSON, U., 'Full-scale/Bench-scale Correlation's of Wall and ceiling Linings' Chapter 13 in: *Heat Release in Fires* Ed: V. Babrauskas & S. Grayson, Interscience Communications London, pp461–477, 1992.
26. BABRAUSKAS, V. *et al.* 'The cone calorimeter used for predictions of the full-scale burning behaviour of upholstered furniture'. *Fire and Materials* **21**, 95–105, 1997.
27. BABRAUSKAS, V. and KRASNY, J. F., *Fire Behaviour of Upholstered Furniture, NBS Monograph 173*, US National Bureau of Standards, Gaithersburg, 1985.
28. *Heat Release in Fires*, Babrauskas, V and Grayson, S J, Editors, Elsevier, London, 1992.
29. ENRIGHT, P. A. and FLEISCHMANN, C. M., 'EC-CBUF Model I Applied to Exemplary New Zealand Furniture' *Proceeding of: Sixth International Fire Safety Symposium*, 147–158, 1999.
30. ENRIGHT, P. A., FLEISCHMANN, C. M., and VANDEVELDE, P., 'CBUF Model II Applied to Exemplary NZ Furniture (NZ-CBUF)', *Fire and Materials*, **25**, 203–207, 2001.
31. TECHNICAL BULLETIN 121, *Flammability Test Procedure for Mattresses for Use in High Risk Occupancies*, California Bureau of Home Furnishings and Thermal Insulation, North Highlands, CA, USA, 1980.
32. TECHNICAL BULLETIN 129, *Flammability Test Procedure for Mattresses for Use in Public Buildings*, California Bureau of Home Furnishings and Thermal Insulation, North Highlands, CA, USA, 1992.

11
Innovative textiles for seating

A. BÜSGEN,
Niederrhein University of Applied Sciences, Germany

Abstract: Motivated by the high potential for innovation in seating, many interesting ideas are leading to new textile materials. Comfort, functionality and design are the main areas of future seating concepts. Spacer fabrics used instead of foam materials can provide humidity transportation to seats; textile integrated elements such as illumination, signs, input devices and data transport devices add functions to cover fabrics of seats; and 3D shaped textiles lead to seamless cover fabrics, avoiding pattern distortions.

Key words: spacer, smart, functional, 3D, textile, fabric.

11.1 Introduction

Textile materials of today for seating in vehicles such as cars and airplanes are mainly used as cover materials (Malvicino and Marzorati, 2003). In addition to aesthetic and design aspects, they offer a high standard of quality. They have to present comfort and safety to occupants, they should be easy to clean and be manufactured using recyclable materials (Fuchs *et al.*, 2003). To ensure a long lifetime, seating textile materials need hard-wearing surfaces and therefore they need to fulfil well-defined technical requirements such as the following (Terraneo, 2005):

- abrasion resistance, e.g. 50000 Martindale cycles (ISO 12947-1-2-3-4)
- pilling strength, e.g. 5 (ISO 12945-2)
- seam strength, e.g. 100 N lengthwise and crosswise
- Tensile strength, e.g. 750 N (ISO 13934/1)

The main markets of textiles for seating are car and airplane seats. There are some important differences concerning the use of car and of airplane seats. Low weight may be important for both, but the weight requirement for airplanes is a good deal more important than for cars because of two reasons: first, weight reduction leads to higher cost savings 'in use' of airplanes and second, in airplanes the number of seats carried is much higher than in cars so that weight saving of one seat is multiplied for airplanes much more than for cars.

Furthermore, car seats will be used mostly by the same person; airplane seats will be used each time by another person (Büsgen, 2007). Consequently, airplane seats have to be cleaned more often, their construction has to cover a wide range of sizes, shapes and weights of occupants, and their design has to meet common taste. In addition, the occupant safety equipment is different in cars and airplanes and modern textiles have to take these safety requirements into account.

11.2 Potential of innovative textiles for seating

Cabins and their interiors are becoming more and more important. According to recent trends, people are spending more time in cars and airplanes and, for this reason, manufacturers like to build or simulate an atmosphere in the cabin which is close to a living room.

Consequently, there is a high potential for innovation in seating and many interesting ideas are leading to new textile materials for this purpose. Comfort, functionality and design are the main areas of future seating concepts. Textiles can contribute to these developments with the following items:

- integrated cooling and heating of seats
- humidity transportation
- softness of seats achieved with recyclable materials
- textile-based integration of functions such as illumination, signs, switches, keyboards, transportation of data and energy supply
- 3D shaped textiles, e.g. for seamless covers of seats, without local stretch and pattern distortion.

Many research and development projects have been undertaken to create spacer fabrics. According to their results, climatisation of seats can be achieved by recyclable spacer fabrics and, even if there are some technical challenges still to be solved, first applications of these textiles have actually entered the market.

Smart textiles are the focus of automotive interior designers for two reasons. First, existing functional elements such as switches and signals can be manufactured in textile materials to incorporate these elements much better into future basic textile covers. Second, additional tasks such as energy supply and data transportation, as well as new aesthetic aspects such as new illumination concepts, can be realised by new and multifunctional textile materials. A nanomaterial-based finish can improve cleaning performance and reduce dust attraction of smart textile surfaces.

3D shaped textiles may solve the task of covering seat geometry with one single piece. Local stretch of fabrics causes differences of fabric densities which lead to different wear properties and pattern distortions. Shaping of

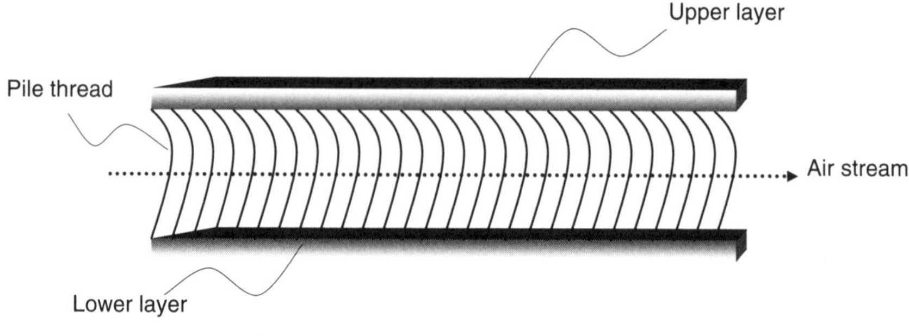

11.1 Spacer fabric.

textiles can be done directly on specialised 3D weaving or 3D knitting machines and may replace draping work on the 2D textiles used until now.

The following chapters will report upon innovations and developments in spacer fabrics, smart textiles and 3D fabrics to be used in seat applications.

11.3 Spacer fabrics for climatisation and comfortable softness

The idea of a spacer fabric is two-fold. First, such a textile material made of a recyclable material (e.g. polyester) can provide a seat with comfortable softness and replace foam layers used up to now, which are much more difficult to recycle because foam is not thermoplastic and can therefore be recycled only by burning (to generate energy). Second, a space between the two layers offers possibilities for climate control of the seat surface by using air streams to heat, to cool or to transport humidity between the upper and lower layers (Fig. 11.1).

Spacer fabrics are made from velvet and plush fabrics produced with weaving and knitting machines and R&D efforts are focussed upon the following methods:

- double jersey weft knitting
- double jersey warp knitting
- double cloth velvet, plush or carpet weaving
- stitch bonded non-wovens
- woven surfaces with integrated pipes

11.3.1 Double jersey spacer weft knitting

Circular weft knitting machines can be used for this type of spacer fabric. Double jersey weft knitting uses two sets of needles. One set is installed at the cylinder and the second set is installed at the rip wheel. The thick-

11.2 Spacer weft knitted seat cover (Trisit, Ulm/Germany).

ness of this spacer fabric is created by the distance between the cylinder and rip needles. At present, this distance is limited to 6 mm, which does not offer enough comfort and softness for seat applications. However, it is suggested that the distance between layers in future should be increased up to 14 mm (Fuchs *et al.*, 2003) to increase softness and comfort. For this, a change in the weft knitting process would be needed to achieve additional control of the cylinder needles.

Another possibility is the use of double jersey flat bed weft knitting machines. According to manufacturers, up to 10 mm thickness can be attained. An advantage of this method is the use of the pattern possibilities of flat bed knitting machines to design the fabric face according to the seat surface requirements. Seat surface and spacer fabric can be combined into one piece in this way (Fig. 11.2).

11.3.2 Double jersey warp knitting

The double jersey warp knitting process has played a major role in the development of spacer fabrics. This type of knitting process is the only knitting technology that is able to create a thickness in the magnitude of 60 mm, which may be the reason why market entry of double jersey warp knitted spacer fabrics has already taken place in the automotive industry.

The distance between the upper and the lower layer of a warp knitted spacer fabric is created by the distance between the knock-over bars. In a double jersey warp knitting machine, a front stitch-needlebar using a front knock-over bar is arranged opposite to a rear stitch-needlebar using a rear knock-over bar. In the past, the distance of these knock-over bars has been limited to 20 mm. However, to achieve the required softness comfort of a seat with a single spacer fabric, a thickness of more than 20 mm is

11.3 Polyester monofil pile threads of a double jersey warp knitting.

necessary. A new development has led to a Raschel machine, called 'High-Distance', which can open the gap between the rear and front knock-over bars far more and produce more voluminous 3D warpknits (Anon., 2009b). This type of fabric makes open surfaces like nets and has to be combined with a cover fabric to create a suitable face for the seat.

The surface layers – fabric face and back – are frequently produced with filament polyester yarns, while the pile thread material uses a monofil polyester (Fig. 11.3). An advantage of the double jersey Raschel machine is the versatile use of different yarn counts and materials. This enables the adjustment of softness parameters to meet the seat requirements.

Attention has to be paid to the pressure distribution of spacer fabrics. Sitting down creates an area load and may lead to abrupt bending of pile threads to one predetermined side. Calculation and computer aided simulation has been carried out to obtain information about the angles and distribution of pile threads (Helbig, 2006), which is necessary to achieve a soft and comfortable pressure bearing area.

The properties of double jersey warp knitted spacer fabrics are favourable for seat applications. Surfaces can be designed that are air permeable, and the distribution of pile threads can be constructed to enable air to circulate freely according to climate requirements. Manufacturers of warp knitted spacer fabrics promise an excellent air distribution, adjustment to every contour, a soft cushioning effect and excellent recovery. In addition, the weight is low compared with other materials and shock absorption is high. Finally, this type of spacer fabric material can be produced in shapes to suit the contours of end components.

The first applications of warp knitted spacer fabrics was for commercial vehicles and special edition cars. 'Maximo EVOLUTION' is an example of a commercial vehicle application (Anon., 2009 b). Active climate control is supported by spacer fabrics, which remove body moisture to avoid damp clothes and seats, and, in addition, can be used to supply heat on cold days. Upholstery pads of spacer fabrics are used in cars such as the Audi A6 (Helbig *et al.*, 2007). Other applications are spacer mesh fabrics for Buick car seat covers, promoted to be the last word in comfort and ease.

11.3.3 Double cloth velvet, plush or carpet weaving

Most machines for this method use two weft insertion systems simultaneously, weaving two layers at the same time with adjustable distance from each other. Weaving machines have to process three different warp systems: upper ground warp, lower ground warp and pile warp systems. The upper and lower warp systems are arranged to build two layers. They open two sheds at a time, one upon another. The third warp system is the pile warp system. Pile warp-ends are controlled to move between the two sheds. They interlace alternating with the upper and lower warp systems. The length of pile warp-ends between the two layers determines the distance of these layers. Consequently pile warp systems need a three-position shedding machine.

In most cases, harness frames of ground warp-ends move heddles with two eyes up and down, controlling two shed formations, one shed of upper warp-ends and one shed of lower warp-ends. The two sheds are opened at the same time and two weft insertions take place at the same time as well, building two woven layers with a distance from each other. This space is filled by the pile warp system, which is controlled mostly by the first two or four harness frames. There is a wide range of variation in the design of the connection between the pile warp and the upper or lower layer. The angle of pile threads, distance of pile threads, intensity of connection between pile warp and ground layers and separation distance between ground layers can be controlled by the machine set up.

The main difference between spacer fabric weaving and velvet, plush or carpet weaving is the elimination of the cutting devices that originally separate the two layers. This, however, creates a challenge to take up the 'sandwich fabric', a process where the spacer fabric becomes badly compressed in obtaining sufficient grip during take-up. Finishing or a wet after-treatment helps to straighten up the pile threads again. A special take-up for spacer fabrics has been developed as a prototype to control the take-up speed of ground and pile warp systems separately, e.g. for weft density variations (Badawi, 2007).

Some properties of woven spacer fabrics are beneficial for use as a foam-replacing seat cover with active climate control. The range of thickness can be adjusted between about 10 to 100 mm (van de Wiele, 1993). This enables one to design the required softness of the seats. Additionally, local softness differences can be achieved by different pile thread distances across the width of a woven spacer fabric. Surfaces can be woven with a wide range of materials and patterns for design and, also, surface permeability can be adjusted according to climatisation requirements of the seats. However, to balance the well-known excellent hardwearing properties of woven surfaces and the required permeability for climate control of the seats, more research and development work is necessary in the future.

Other properties may be undesirable for seat applications. In general, woven materials offer comparatively low elasticity leading to poor mouldability or poor drapeability when the fabric has to cover a shaped geometry such as a seat. Unlike ordinary woven fabrics, cutting and sewing of woven spacer fabrics is more time- and cost-intensive.

11.3.4 Stitch bonded non-wovens (Multiknit)

Multiknit is a stitch bonding process using 100% carded fibres with lengthwise orientation. A two-stage stitch bonding creates a spacer fabric, where pile fibres maintain a distance between the two surface layers of 2 to 16 mm. To produce a cushioning effect, a thermal setting is necessary.

Applications of this spacer fabric will probably be limited to the cushions of seat upholstery, which are recyclable and may replace foam cushions. A first presentation by a car component supplier (Anon., 2009a) shows that there is a market potential for this kind of fabric.

Airplane seats are another interesting application shown to the public. A non-woven spacer fabric is used to ventilate the humidity of a human body sitting for many hours on a plane seat. The *Klimovation* seat of the automotive supplier Car Trim, gets by without an electrical ventilation unit and therefore saves weight in the range of 1.4 kg per seat.

11.3.5 Woven surfaces with integrated pipes

A new and little-known alternative to two-layer fabrics with a pile thread or fibre distance system is a weaving process of single layer fabrics with integrated pipes for air or liquid transportation. The fabric has two layers, of which one is longer and is recoiled to a controlled extend by the beat-up of the reed. Intermediate interlock of the recoiled length leads to pipes up to a height of 10 mm (Fig. 11.4).

Fabrics of this kind can be woven on modern weaving machines. The wear resistance of the warp material used is important for creating sufficient recoiling length and hence pipe height. Air permeability of pipes can be adjusted by weave construction. Supporting air ventilation through a seat, this fabric may be used as the vertical inner side of a seat, providing climate control and – to a limited extent – softness for the passenger.

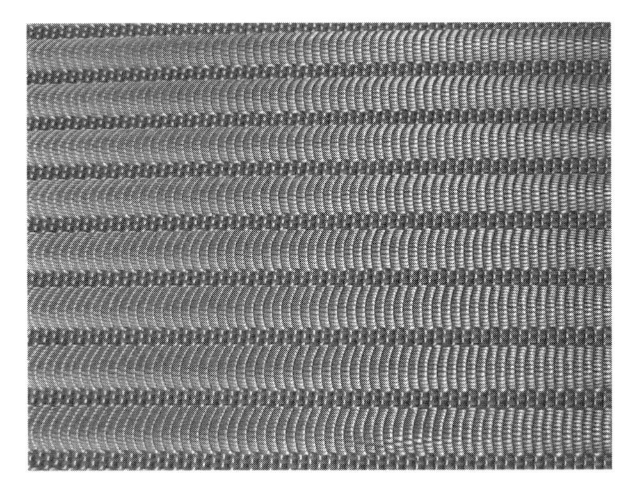

11.4 Woven polyester fabric with integrated pipes.

11.4 Smart textiles for seating

Fashion and design have discovered the potential of smart textiles long ago. Until today, a mass market has not been established in this area even though a huge number of fashion shows and publications has promoted the ideas of smart textiles. Technical applications, in contrast, seem to offer a higher potential for smart textiles, and car interior textiles such as seat fabrics may play an important role for this market in future.

The integration of light-emitting elements into a car seat fabric can help one find switches such as for electronic seat adjustment, may illuminate dark areas (e.g. part of the interior floor around the seat) or can display seat adjustments such as heating or cooling control functions of the seat. For plane seats, it would be interesting to integrate displays into the backs of seats for signals or announcements to passengers.

Basically, there are three different possibilities for illuminating seat fabrics:

- LEDs
- electroluminescent fibres
- optical fibres

LEDs offer a very high luminance (luminous intensity per area), up to 100 000 cd/m^2 (Table 11.1). They are useful mainly for punctuated illuminations and they are not damaged by water due to their low voltage and inner resistance (Tillmanns *et al.*, 2008). As one of the first applications in cars, Rolls-Royce's Phantom Coupe, presented in March 2008 to the public, uses hundreds of LEDs integrated into the roof fabric to create a dazzling illusion of the sky at night.

Table 11.1 Luminance of fabric elements (Tillmanns *et al.*, 2008)

Illuminating medium	Luminance (cd/m^2)
LED	~100000
Electroluminescent fibre	~100
Optical fibre	<50000 dependent on the angle between direction of beam and line of sight

Illumination of lines and areas in transportation vehicles can be achieved with electroluminescent materials. They are available as high voltage cables with mechanical properties similar to coarse monofilaments. Weaving these materials directly into a seat fabric requires a protective coating against friction damage by, for example, reed dents. Furthermore, a large bending radius of 15 to 20 mm has to be provided; smaller radii due to weave constructions may cause permanent damage to the material.

A focus of research and development for textile illumination is upon optical fibres (e.g. polymethyl methacylate fibres). Light emission, however, is relatively low compared with electroluminescent cables. Optical fibres need a light source, which is most often an LED, feeding light beams into the longitudinal axis of the optical fibres. Light emission is achieved by scarification of the fibre surface or by micro reflectors, distributed inside the fibre. From the light source to the end of the fibre, the emitted light intensity decreases visibly. A benefit for future seat application is that optical fibres can be small and soft, that they do not change the fabric character if distributed detached inside the fabric, and that they are robust to humidity and cleaning.

Sensoric textiles offer another possibility for adding functions to seats. First, textile keyboards or keypads that are part of the seat cover textile can replace the, up to now used, switches and buttons made of plastic materials. Second, textile sensors can determine actual occupation of seats. In future they could measure weight, size or position of a seat occupant and control information with this, e.g. individually adapted activation of airbags.

Integration of keypads or keyboards into seat cover fabrics can be achieved by various methods:

(i) printing and coating with a conductive paste
(ii) embroidering a conductive yarn onto the fabric surface
(iii) designing a two-layer fabric with conductive yarns in both layers
(iv) designing a fabric having small two-layer bulges

Methods (i) and (ii) are useful for single-layer seat textiles. Both are easy to clean and proven in applications for smart clothes (Post *et al.*, 2000;

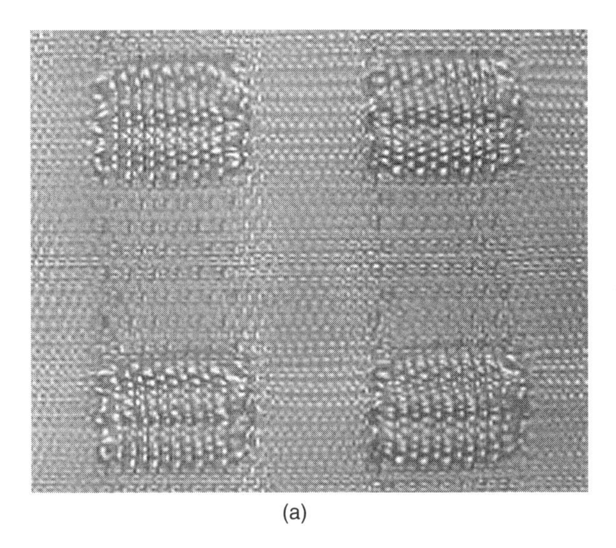

(a)

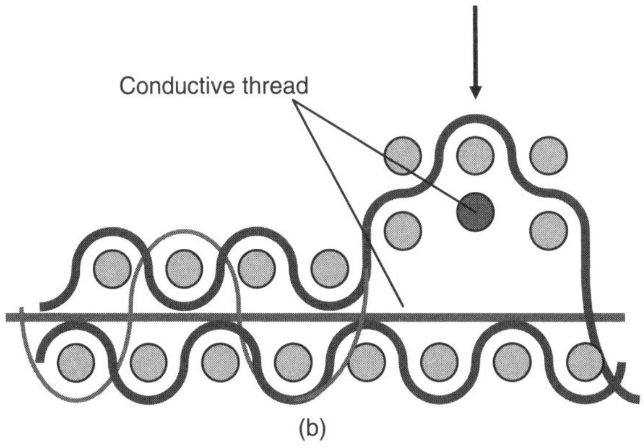

(b)

11.5 Woven bulges for buttons (a) and weave construction thereof (b) (Büsgen, 2007).

Meoli and May-Plumlee, 2002). Sensing can be achieved, e.g. by measuring the capacity of conductive surface areas, which changes when touched by even a finger.

For a two-layer fabric, Method (iii) is more appropriate. If spacer fabrics are used, a basic two-layer fabric is already at hand. Recovery of the pressed down space is obtained by pile threads, so a time-limited contact of conductive yarns between upper and lower ground layers can be designed. This method depends on the successful market entry of spacer fabrics, but if this precondition is fulfilled, it will be a cost-effective possibility for textile keyboard integration in future.

Method (iv) offers a tactile aid to a seat passenger if buttons are at a position where they are not in sight. As an example, Figures 11.5 and 11.6

11.6 Woven bulges for buttons with 10 mm height (Büsgen, 2007).

demonstrate realisations of this method with weaving technology (Büsgen, 2007). The Weaves may be constructed on ordinary weaving machines. Bulges are woven by creation of local recoiling zones of one layer, where two layers are not connected to each other. After reconnection, the recoiling layer builds a bulge and – if conductive yarns are used in both layers – this can be used as a button. An advantage of this method is that a given weave design of a seat cover could be integrated into the bulging button formation.

Some of these ideas have been developed into products that are ready to enter the automotive market. First examples of textile key buttons and illuminated elements integrated into a seat are now offered by seat manufacturers (Anon., 2006). As an example, a textile reading lamp with textile on/off-switches was presented at the Mtex-Chemnitz Fair in 2006.

11.5 Seamless 3D shaped fabrics for seating

Applying a fabric to a seat requires draping, local stretching and tensioning of fabric materials. Regardless of the manufacturing method, a pattern distortion is inevitable in many cases. A fabric that already had the geometry of a seat could avoid such pattern distortions. If a 3D fabric is made as a one-piece shell, application is easier and consequently manufacturing costs can be saved. Another interesting aspect is that 3D fabrics are individually designed for a specified application. If the seat geometry is to be attained, individual requirements such as local reinforcements can be taken into account, to improve the quality or lifetime of the seat cover. Furthermore, many ideas for the smart seat fabrics described earlier, can be realised more easily by a seamless 3D fabric because functional yarns, e.g. for light emission or data transportation, can be incorporated into the fabric uncut. Finally, a combination of 3D fabrics and smart fabrics for seats could lower the manufacturing costs as the integration of smart yarns can be achieved directly by a fabric production process and does not require additional working or processing steps.

3D weft knitting and 3D shape weaving are the two basic processes upon which discussion about 3D seating fabrics is focussed.

11.5.1 3D weft knitted shapes

3D weft knitting was developed first. Work on shaping weft knitted fabrics directly at the machine has been done to develop seamless garments and clothing products. New devices and additions to flat bed weft knitting machines have led to the so called 'knit and wear' process of Stoll/Germany or the 'whole garment' process of Shima Seiki/Japan. This may have given the basic idea of a weft knitting process for seamless knitted 3D seat fabrics. In 1985, Courtaulds and General Motors developed the first 3D knitting process to manufacture seat covers (Sorge, 1994).

Shape is created during flat bed weft knitting by additional and local weft thread processing actions. In general, knitting needles are controlled and activated or deactivated, building local stitch courses of limited length. According to the intensity of added local stitch courses, a bulging zone is created. Because a 3D shape sometimes needs only few stitches at a local stitch course, requiring a cam running across the whole width of a fabric, production time is increased compared to ordinary weft knitting.

This technology has also been used to develop seamless weft knitted 3D reinforcements for helmets. Seamless weft knitted seat covers have been developed by Trisit/Germany and under the tradename *Teknit* by Camira Fabrics Ltd/UK. Both companies use Stoll's programmable flat bed weft knitting machines, e.g. the CMS5/30 type, and add sophisticated machine control processes to achieve 3D shapes (Legner, 2001, 2003).

A disadvantage discussed has been the wear resistance of weft knitted fabrics. In comparison to woven seat cover materials, weft knitted fabrics sometimes show shorter lifetimes and therefore the 3D weft knitting process may be more useful for seats that are not exposed to very intensive friction. Advantages of this method are waste reduction compared to 2D tailored seat covers and the broad variety of patterns given by flat bed knitting machines.

11.5.2 3D woven shapes

The other process used to manufacture 3D shaped seat fabrics is a 3D weaving process called 'shape weaving'. This process was initially developed to make seamless reinforcement fabrics for composites. Further research has led to other areas of application such as medical textiles, clothes, hats, car interior textiles and seat cover materials.

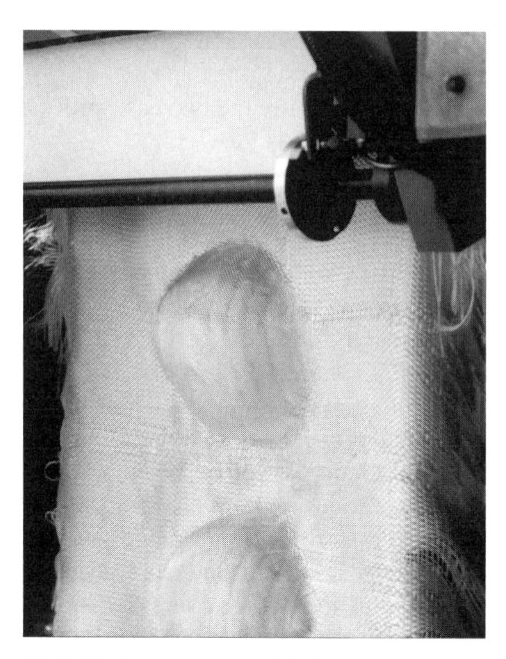

11.7 3D shaped woven fabric.

Producing 3D shaped geometries at a weaving machine requires control of local thread spacing for weft and for warp yarns. If the distance of yarns in a defined zone of the fabric is increased, this zone bulges up into the third dimension according to the local area increase (Fig. 11.7). A shape weaving machine has to use devices to position yarns during interlacing, e.g. to spread yarns locally or to take up different lengths locally. Additional filling threads can be filled locally into spaces between increased threads, e.g. to keep fabric areal weight even across a 3D surface or to increase reinforcement locally as required.

3D fabrics woven according to the shape weaving process result in a weave construction that is different from 2D fabrics. Thread orientations and spacing continuously change according to woven shape and specific thread control during the 3D weaving, as demonstrated in Fig. 11.8 (Büsgen, 2008). Consequently, the weave has to be adapted to changing local conditions of the shaped fabric, which leads to a huge amount of control data. To handle this amount of data, CAD software for shaped woven fabrics has been developed (Birghan *et al.*, 2007 a–c). This software enables a simulation of weave construction and is used for design and quality optimisation of shaped woven fabrics.

First applications of 3D woven fabric shapes have been developed for car interior textiles. As an example, a 3D woven cover fabric was made to integrate smart textile functions such as switches and illuminating elements

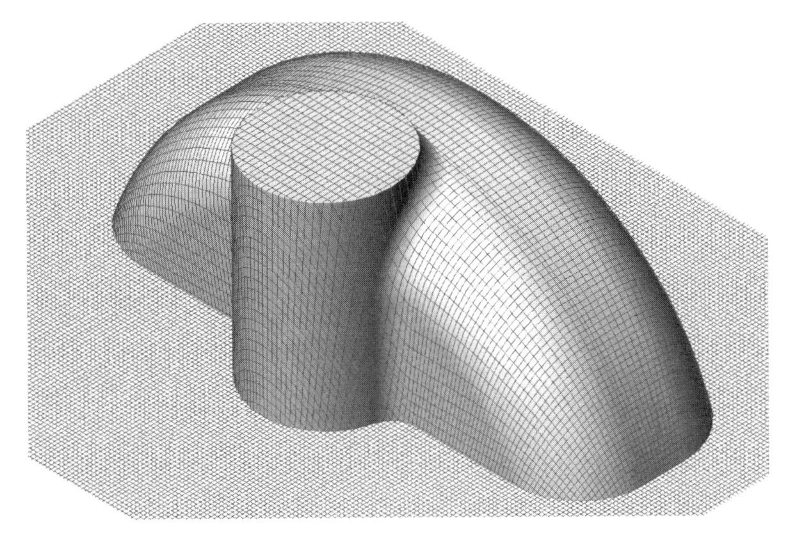

11.8 Thread orientation and spacing of a 3D woven fabric (Büsgen, 2008).

to prove that 3D weaving is able to combine smart textile and 3D weaving technology in a cost effective way. It remains to be seen if this method will enter automotive or aircraft markets.

11.6 Sources of further information and advice

Further information is available from machine manufacturers and at research institutes working on modern textile technology, including interior textiles of vehicles, e.g.:

KARL MAYER Textilmaschinenfabrik GmbH
Brühlstraße 25
63179 Obertshausen, Germany
(Publication: Kettenwirkpraxis, a quarterly technical journal)

H. Stoll GmbH & Co. KG
Stollweg 1
72760 Reutlingen, Germany
TU Dresden

Fakultät Maschinenwesen
Institut für Textil- und Bekleidungstechnik
01062 Dresden, Germany

Trisit Textiltechnologie GmbH & Co. KG
Im Lehrer Feld 9
89081 Ulm, Germany

Car Trim GmbH
Friesenweg 19
08529 Plauen, Germany

Institute of Textile and Clothing at Niederrhein University of Applied Sciences
Webschulstr. 31
41065 Mönchengladbach, Germany

Shape 3 Innovative Textiltechnik GmbH
Friedrich-Engels-Allee 161
42285 Wuppertal, Germany

11.7 References

ANON. (2006), *Textiles with a control function*, Mtex Newsletter, No.2/2006, pp.2

ANON. (2009a), *More levels of freedom/Autoland Saxony*, pp. 57, available at: www.autoland-sachsen.com/Inhalt_1_2008/56_57_Car_trim.pdf [Accessed 12.1.2009]

ANON. (2009b), *Maximo evolution active*, available at: http://www.grammer.com/english/lm_maximo_evolution_active/ [Accessed 12.1.2009]

BADAWI, S. (2007), *Development of the weaving machine and 3D woven spacer fabric structures for lightweight composite materials*, Ph. D., Dresden Technical University, Dresden/Germany

BIRGHAN, A., TILLMANNS, A., FINSTERBUSCH, K., BÜSGEN, A. (2007a), 'Simulation and calculation of seamless woven 3D shells (Part 1)', *Technical Textiles*, 02/2007, pp E144 f.

BIRGHAN, A., TILLMANNS, A., FINSTERBUSCH, K., BÜSGEN, A. (2007b), 'Simulation and calculation of seamless woven 3D shells (Part 2)', *Technical Textiles*, 03/2007, pp E180 f.

BIRGHAN, A., TILLMANNS, A., FINSTERBUSCH, K., BÜSGEN, A. (2007c), 'Simulation and calculation of seamless woven 3D shells (Part 3)', *Technical Textiles*, 04/2007, pp E266 f.

BÜSGEN, A. (2007), *New textiles for seats: Workshop*, IQPC Conference 'Innovative seating', Frankfurt a.M., Germany, 6. Feb.

BÜSGEN, A. (2008), *Simulation and realisation of 3D woven fabrics for automotive applications*, 1. World Conference on 3D Fabrics, Manchester, 10–11. April

FUCHS, H., SCHILDE, W., ERTH, H. (2003), *3-D automotive textiles – A comparative evaluation*, 42nd international Man-made Fibres Congress, Dornbirn/Austria, 17–19 September

HELBIG, F. (2006), *Gestaltungsmerkmale und mechanische Eigenschaften druck-elastischer Abstandsgewirke*, Ph.D. Chemnitz, University of Chemnitz/Germany – Faculty of Mechanical Engineering

HELBIG, F., HEINRICH, H.-J., SCHULZE, D., BREDEMEYER, J. (2007), *New dimensions for upholstery*, 14th International Techtextil-Symposium for Technical Textiles, TT.6.6, Frankfurt a.M., Germany, 14–17. June

LEGNER, M. (2001), 'New approaches to the design of automobile interiors using three-dimensionally knitted fabrics', *Melliand International*, Vol. 7, June, pp. 127–129

LEGNER, M. (2003), '3D products for fashion and technical textile applications from flat knitting machines', *Melliand International*, Vol. 9, September, pp. 238–241

MALVICINO, C., MARZORATI, D. (2003), *Textile applications in the automotive industry,* 42nd International Man-made Fibres Congress, Dornbirn/Austria, 17–19

MEOLI, D., MAY-PLUMLEE, T. (2002), 'Interactive electronic textile development', *Journal of the Textile and Apparel, Technology and Management (JTATM)*, Vol 2, Issue 2, NC State University

POST, E., ORTH, M., RUSSO, P., GERSHENFELD, M. (2000), 'E-broidery – Design and fabrication of textile based computing', *IBM Systems Journal*, Vol 39, Nos 3&4, pp. 840–860

SORGE, M. (1994), A stitch in time: Inland Fisher guides low cost, high tech seat covers, *Ward's Auto World*, pp. 52

TERRANEO, L. (2005), 'Innovative suede: Deluxe appeal and care for the nature', *Automotive Interior*, 6.7.2005

TILLMANNS, A., HEIMLICH, F., BRÜCKEN, A., BÜSGEN, A., WEBER, M. (2008), 'Investigation of the possibilities to realize light effects in textile fabrics', *Technical Textiles*, 4, pp. 182–183

VAN DE WIELE, M. (1993), *Production and applications of woven spacer and technical textiles*, 7th Weaving Colloquium, ITV Denkendorf

Index